The
FARM SAFETY
Handbook

The
FARM SAFETY
Handbook

Rick Kubik

Voyageur
Press

First published in 2006 by Voyageur Press, an imprint of MBI Publishing Company, Galtier Plaza, Suite 200, 380 Jackson Street, St. Paul, MN 55101-3885 USA

MBI Publishing Company titles are also available at discounts in bulk quantity for industrial or sales-promotional use. For details write to Special Sales Manager at MBI Publishing Company, Galtier Plaza, Suite 200, 380 Jackson Street, St. Paul, MN 55101-3885 USA

ISBN-13: 978-0-7603-2384-4
ISBN-10: 0-7603-2385-2

Editor: Amy Glaser
Designer: LeAnn Kuhlmann

Printed in China

About the author

Rick Kubik is a certified crop adviser for the American Society of Agronomy and lives in Calgary, Alberta, Canada.

On the front cover, main image: Traversing a slope can cause a tractor to tip sideways. Factors that suddenly increase the danger of tipping include a downhill wheel falling into a pothole or animal burrow, an uphill wheel jarred upward by a stone or dirt pile, or an implement that lifted up and raises the overall center of gravity.

Detail image: PTO shaft guards are designed to spin loosely on the driveshaft so that if a person or object accidentally touches a moving shaft, the guard will stop. This helps prevent entanglement around the shaft or in the flexible joint. The yellow shaft in the foreground has the complete guard in place.

On the back cover: Mowing operations near public roads involve the risk to others as rocks or garbage are ejected at high speed and strike passing cars. For this reason, a flail-type mower may be preferred over the convetional rotary mower.

CONTENTS

INTRODUCTION
THE OFFGUARD PROJECT

To help in your review of farm safety hazards and responses, this book looks at what to watch out for in typical farming situations and provides case studies of incidents that have occurred in these situations. In many cases, that takes the form of federal and state fatality reports. As valuable as these are, the victims, unfortunately, cannot speak about their involvement and lessons learned.

That's why these fatality reports are supplemented by six personal stories of the events and consequences of farm safety incidents where the victims survived to tell the tale. These stories were originally part of a photographic exhibition and catalog called *OffGuard: Farmers and Machinery Injuries*. This project was the result of an innovative collaboration of partners brought together by the Institute of Agricultural Rural and Environmental Health (formerly Centre for Agricultural Medicine) at the University of Saskatchewan, Canada.

Farm injuries have identifiable patterns. When patterns can be identified, an injury is predictable. By eliminating risk factors, the injury is preventable.

Through the vision and efforts of all of the participants in this project, partnerships were assembled that linked the Centre for Agricultural Medicine with the Department of Art and Art History, the Kenderdine Art Gallery, and the Photographers Gallery. While each partner played an invaluable role in the realization of the exhibition, catalog, and Provincial tour, the ultimate goal of the project was to develop an educational tool that could reach those most directly affected by the exhibition content—the rural population. Communicating a message of prevention is of primary importance to arrest further tragedies of this nature. To allow a lasting visual and textual record to be compiled in the compassionate service of others is a generous act.

When a farmer is injured, the event is not usually witnessed or investigated, and is often underreported. Radio or newspaper coverage, when it exists, focuses on the outcome of the incident and provides sketchy, if any, details about what happened. The injured person and family, concerned that others in the community may view the incident as incompetence, understandably create a veil of silence around the event. Burdened with grief and guilt, they may take solace in regarding the injury as an accident, a freak event, or even an act of God. Aware of the family's pain and considerate of their privacy, outsiders respect their silence. Privately, outsiders may attribute the incident to carelessness and reassure themselves that such a thing could never happen to them. A year later, the only people to have learned from the incident are those who were directly involved.

Contrast this with the situation in other hazardous industries such as mining, forestry, and manufacturing. Injuries there are almost always witnessed by one or more coworkers who see the injured person's pain and suffering firsthand. When the individual returns to work following a serious injury, coworkers observe the daily struggle of working with a permanent disability. The injured person, witnesses, and coworkers participate in an accident investigation that is not intended to lay blame but attempts to identify the immediate cause, contributing factors, and basic cause of the incident. Changes are made to equipment, work practices, or the environment to prevent such an incident from recurring. A year later, many people have learned from the incident and the workplace is safer as a result.

We need to learn about safe work practices, make the decision to adopt them, and then use them consistently. This may require changing patterns of behavior that have existed for generations.

The science of injury research is much more than the collection of facts and figures. Researchers use cumulative data and detailed reports of incidents to develop a clear understanding of what causes injuries. The investigation, tracking, and reporting of incidents has helped reduce injuries in modern industry. Workers Compensation Board statistics identify high-hazard jobs and activities, and insurance costs based directly on injury rates provide employers with an incentive to improve workplace safety.

Only recently has the science of injury research been available in the agricultural setting. Recurring patterns of farm work–related death and injury across the country and throughout Saskatchewan reveal that agricultural deaths and injuries are not random or isolated accidents. Farm injuries have identifiable patterns. When patterns can be identified, an injury is predictable. By eliminating risk factors, the injury is preventable.

There is no easy answer to the question of how farm injuries can be prevented. Injuries frequently involve more than one risk factor. Some risk factors, such as weather and machinery design, can't be controlled by the farmers. But farmers can take control of their own safety by making the decision to adopt safe work practices and use them consistently. For example, research has shown that starting a tractor from the ground is a major risk factor for operator runover. The related safe work practice is to start the tractor only from the operator's seat. That takes seconds longer than ground-starting, but the time saved is insignificant when the benefit is weighed against the risk. One in four tractor runovers due to ground-starting results in death, and the remaining three result in serious and often permanent disabling injuries. We need to learn about safe work practices, make the decision to adopt them, and then use them consistently. This may require changing patterns of behavior that have existed for generations.

Change isn't easy, but the ability to change is essential for survival. Immerse yourself in the images of these farm injury survivors and listen to what they have to tell you. They are prepared to share what they have learned so others will not be caught off-guard by a farm machinery injury.

The OffGuard photographs were taken by Mark Ballantyne, Paula Reban, and Naomi Friesen as course work under the supervision of Professor Brenda Pelkey. The photos are accompanied by explanatory text written by Julie Bidwell, which was based on interviews with the subjects conducted by Bidwell and College of Nursing graduate students Roxanna Kaminski and Hope Bilinski. Statistical information provided for the exhibit was based on "Fatal and Hospitalized Farm Injuries in Saskatchewan 1990–1996," which was published in June 2000 by the Saskatchewan Farm Injury Surveillance Program in collaboration with the Canadian Agricultural Injury Surveillance Program. A copy of the report may be obtained by contacting the Institute of Agricultural Rural and Environmental Health.

OffGuard Project Partners:
• **Institute of Agricultural Rural and Environmental Health: University of Saskatchewan, Saskatoon, Saskatchewan**
    Helen McDuffie, manager,
        Rural Health Extension Program
    Julie Bidwell, occupational health nurse,
        Rural Health Extension Program

• **College of Arts and Sciences: University of Saskatchewan, Saskatoon, Saskatchewan**
    Brenda Pelkey, professor,
        Art and Art History Department
    Mark Ballantyne, Paula Reban, and Naomi Friesen,
        fourth-year photography students

• **College of Nursing: University of Saskatchewan, Saskatoon, Saskatchewan**
    Donna Rennie and Karen Semchuk, professors
    Hope Bilinski and Roxanna Kaminski,
        graduate students

• **Kenderdine Art Gallery: University of Saskatchewan, Saskatoon, Saskatchewan**
    Kent Archer, curator
    Betsy Rosenwald, catalogue designer

• **The Photographers Gallery: Saskatoon, Saskatchewan**
    Donna Jones, director

• **Farmers with Disabilities Program, Saskatchewan Abilities Council: Saskatoon, Saskatchewan**
    Bob Elian, coordinator
    Participating program members

• **Saskatchewan Farm Injury Surveillance Program, Institute of Agricultural Rural and Environmental Health: University of Saskatchewan, Saskatoon, Saskatchewan**
    Louise Hagel, injury epidemiologist
    Donna Rennie, faculty advisor

The OffGuard Project Partners gratefully acknowledge the financial support of Agriculture and Agri-food Canada, Canadian Coalition for Agricultural Safety and Rural Health, Museums Association of Saskatchewan, Saskatchewan Agriculture and Food, Saskatchewan Lotteries Trust Fund, the Canada Council for the Arts, and the Cyril Capling Trust Fund, College of Agriculture, University of Saskatchewan. This exhibition could not have taken place without the support of the Farmers with Disabilities Program and the generous participation of its members.

A portion of the author's royalties for this book have been donated to support and continue farm safety work at the Institute of Agricultural Rural and Environmental Health.

Note: The original OffGuard exhibition is available for touring for farm safety education purposes. For details please contact:

Institute of Agricultural Rural and Environmental Health
Box 120 RUH, 103 Hospital Drive
Saskatoon, SK S7N 0W8, Canada
Phone (306) 966-8286
Fax (306) 966-8799
E-mail: aghealthandsafety@usask.ca
Webpage http://iareh.usask.ca

CHAPTER 1
SAFETY BASICS

▶ *In any given year, the number of farm fatalities is usually exceeded only by those related to mining.*

▶ *In any given year, more than 700 Americans die in farm-related accidents, with children accounting for between 100 and 300 of those deaths.*

Living and working out in the country may look like a healthy, safe, outdoorsy activity, and for the most part it is. Many farmers live to a ripe old age, as do most people in other risky occupations. However, like any other risky occupation, the farm environment has many hazards you may not know about, or appreciate the full extent of, especially if you are new to the field. Without knowledge of potential safety problems, those hazards can quickly take a finger, a limb, or even your life. Adding to the severity of farm accidents, incidents may occur when the victim is working alone, far from other people who could help rescue or extract the victim, provide first aid, or call emergency services.

The dangers in farming are not likely to disappear anytime soon because of farming's inevitable dealings with problems such as heavy machinery, fast-moving drivelines, and unpredictable animals, all made worse by the stress of being at the mercy of the weather and trying to get things done while favorable weather holds.

What can change much more easily is how you personally deal with those dangers. It's your choice whether you are going to educate yourself about typical dangers others in your situation have encountered or learn the hard way. Once you are aware of what often happens and why, it's your choice on which precautions fit the way you live and work. That's why this book is written to help provide you with practical examples of safety hazards that have been observed on farms; the precautions that can help avoid personal risk, injury, and death; and ways to make your farm a safer environment for yourself, your family, and visitors.

The situations and responses you read in this book can help set the standard of safety-mindedness for others on your farm as well. Taking time to review what

hazards you're likely to encounter and what others have done in similar situations increases your chances of finishing the day with all body parts and functions intact.

BASIC SAFETY TOOLS

Any job needs to start with having the right tools. When it comes to the job of keeping alive and well, it is important to have the necessary basic tools, such as:
- Operator's manuals as a guide to safe operating methods.
- Contact information for emergency services.
- First aid kits.
- Fire extinguishers.
- Accessible, reliable source of high volumes of water for fighting larger fires.
- A plan or a general idea of what to do, where to go, or who to contact in an emergency. In particular, make sure children know what they need to do or where to go.

Along with having the tools, you need to make sure they remain effective and easy to use, not lost, broken, or out-of-date.
- Keep operator's manuals where they are easy to find and use for reference.
- Post emergency contact information prominently near every phone.
- Inspect first aid kits at least every year to make sure they contain sufficient, up-to-date supplies. Keep kits where an injured person can quickly and easily access them.
- Inspect fire extinguishers at least every year and recharge as needed.
- Check flashlights regularly to make sure they will work when needed.

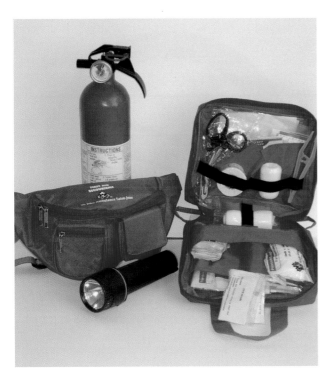

Does your farm have a sign directing ambulances or fire trucks to your location when summoned? Your farm may look distinct to you, but that's not necessarily true for others. A simple, noticeable sign can mean the difference between speedy aid and an agonizing wait for emergency help.

Prevention is the first line of defense against safety-related injury and damage. Back-up your prevention program with a second line of injury- and damage-treatment tools.

MAKE "LEARNING BY ACCIDENT" A THING OF THE PAST

Farm machinery and animals can either help you enjoy rural living and perhaps make a living at it, or they can cause injury, pain, and long-term disability. The difference quite often comes from finding out beforehand the right way to deal with them. Take every opportunity you can to learn about the right way to manage machinery and livestock so you can put that knowledge to use and protect yourself and your family.

When you buy, rent, or borrow equipment, take every opportunity you can get for training that shows you proper and safe usage.

Get the operator's manual and read it. Even for old machinery, manuals are often available from dealers or on Internet sites such as eBay.com, manuals.us, or ytmag.com.

Take a good look at any warning labels on the machine. They are there to alert you to hazards at that particular point on the machine.

If you're acquiring livestock, especially breeding stock such as bulls, find out from the previous owners any special handling techniques that can help keep you from harm.

MAKE TIME FOR MAINTENANCE

Many accidents occur because of broken machines that owners do not have the knowledge or have not taken the time to maintain and repair correctly. Driveline guards are left off, or the operator stands beside the tractor to short-circuit the starter because it will not start the normal way (the number one factor in tractor runover accidents). At a lesser level of injury, a common cause of cut and bruised hands can be not taking the time to spray oil on a rusty bolt. When you are straining on a wrench with all your might trying to turn a rusty bolt, if the bolt suddenly lets go or the wrench slips, you can end up with nasty cuts and bruises when your knuckles slam into hard parts, hence the term "knuckle busters" for open-end wrenches.

• Keep equipment properly lubed, tightened, sharpened, and maintained. Poorly maintained equipment is more likely to expose you to danger when you get close to a problematic mechanism due to lack of maintenance.

• Keep driveline guards in place and properly secured so you are protected from entanglement in chains, belts, and spinning shafts.

• Take a few seconds to block the wheels when working on tractors or implements or when unhitching implements. Unexpected rolling or slipping can lead to nasty consequences.

If something breaks, schedule time for a proper repair as soon as it is practical. Don't rely on baling wire and chewing gum fixes for long. They can keep you going for a while, but they can also fail unexpectedly and increase safety hazards.

A container may explode if exposed to heat or if it has a physical shock, such as being dropped.

STAY IN TOUCH

One of the great attractions of working in the country is the freedom of being on your own to do what you want, when you want, without the need to explain and discuss your plans to others.

On the other hand, if another person has a general idea of where you are and when you expect to be heard from again, they can bring help if you're overdue. Many of the fatal accident reports related to farm safety contain a chilling observation to the effect that the victim was not missed for a long time or was discovered by a passerby. If service coverage is reliable, consider using a cell phone or radio to check in with others now and then, even if only to leave a voice-mail message.

Since cell phone coverage is spotty in many agricultural areas, check into a pocket Global Positioning System (GPS) unit that includes a medical alert button you can hit in an emergency.

A note in a day log kept at a known location, such as your farm workshop bench, lets people know where you've been. Over the years it can make a great historical resource of your life on the farm.

Keep a friendly lookout for neighbors and family members so that you are able to render timely aid if they are in trouble.

Some products are capable of catching fire in the presence of a spark or open flame under normal working conditions. They include gases, liquids, solids, and aerosols.

LEARN THE SIGNS OF DANGER

Some of the hazards of farming are clearly marked with certain symbols on products. The Workplace Hazardous Materials Information System (WHMIS) is an international system of symbols that provides health and safety information about controlled products, such as solvents and flammable materials.

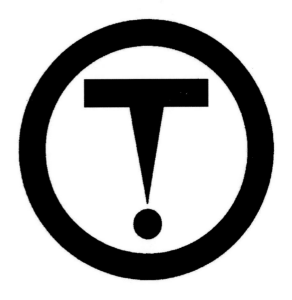

There are products on your farm that may spontaneously create heat or catch fire under normal conditions of use. They may also emit a flammable gas or spontaneously catch fire when in contact with water, or increase the risk of fire if the product comes in contact with flammable or combustible materials.

Life-threatening and serious long-term health problems can be caused by products commonly found on a farm. These products can create less severe but immediate reactions when someone is repeatedly exposed to small amounts.

Some substances can cause immediate injury or death when a person is exposed to small amounts.

Some products contain microorganisms that can cause harmful reactions.

Certain products on the farm can destroy skin or eat through metals.

Substances may self-react dangerously (e.g., explode) when exposed to physical shock or to increased pressure or temperature. They can also emit toxic gases when exposed to water.

When towing with a tractor, hitch only to the drawbar. This ensures the applied load will keep the tractor level front to back.

Do not attach a tow line to any tractor part above the drawbar. The applied load will cause the tractor to rotate around the rear axle and result in a backward overturn.

CASE STUDY: TOW LINE BREAKAGE

1) A 45-year-old male farmer died from injuries sustained when he was struck by a chain being used to pull a piece of equipment that was stuck in a field. One end of a tow chain was hooked to the stuck implement and the other end was connected to a stretchable snap strap (strong, wide, flexible fabric band used for towing), which was connected to the tractor. On one attempt to pull the cultivator free, the chain slipped from where it was hooked to the cultivator. The chain was propelled toward the tractor and the victim, who was seated in the enclosed cab. It shattered the rear and front windows of the tractor cab, struck the victim, and caused him to lose control of the tractor. The tractor traveled several hundred feet and entered a ditch along a public road. The cab door opened and the victim fell from the tractor. He was run over by the left-rear dual wheels. The tractor continued to travel in a large circle until the right-side wheels collided with and drove onto the field cultivator, where it also became stuck.
—from Minnesota FACE Report #96MN04001

2) A 41-year-old male part-time farmer/dentist was crushed when the borrowed 1952 farm tractor he was operating flipped over while he was attempting to pull recently cut sapling stumps (6 to 8 inches high and about 1 1/2 inches in diameter) out of the ground with a log chain. The victim had used the borrowed vintage tractor many times before without incident. In this case, because the rear hitch was missing, the victim secured the log chain around the tractor seat. When power was applied, the tractor flipped over backward (front over back), trapping the victim under the steering wheel and crushing his chest. The victim's spouse witnessed the accident and immediately called 911. A wrecker (tow truck) had to be called to remove the tractor from the victim.
—from Indiana FACE Report #95IN03901

When you see any of these symbols, find out the appropriate measures for working safely in the presence of the marked material. For more complete information, inquire at the place that sold the product, or do an Internet search for WHMIS symbols.

CONSUMER PRODUCT HAZARD SYMBOLS

While WHMIS labels apply to controlled products, labels on consumer products may also have symbols denoting poisonous, corrosive, flammable, and explosive hazards. Take a moment to look at the symbol to assess the level and kind of hazard associated with the product.

Labels for consumer (restricted) products, such as pesticides, have symbols similar to the WHMIS symbols above contained inside an inverted triangle, diamond, or octagon depending upon the degree of hazard.

Symbol	Signal Word	Level of hazard
Octagon	DANGER	High
Diamond	WARNING	Moderate
Triangle	CAUTION	Low

A symbol may also be included inside the octagon, diamond, or triangle hazard symbol. For example, the skull and crossbones symbol inside an octagon with the words "DANGER POISON" indicates a high poison hazard.

LEARNING THE SIGNS OF DANGER
TOWING

Living on a farm often presents situations where vehicles, equipment, and large animals can become stuck in mud, snow, or rough terrain. In addition to dealing with your own troubles, you may be contacted to help motorists who have slid off the road.

Solving the problem is not simply a matter of tugging with any old towing line that happens to be lying around. Parts such as ball hitches, clevises, chains, and complete bumpers can break loose and become dangerous missiles capable of punching right through cabs and windows, creating great danger for those nearby.

If a tow line breaks, its parts can become a deadly projectile. All towing materials—chains, cables, tow straps, and nylon ropes, along with the hooks at their ends—are dangerous when they recoil from a stretched condition. When towing, hitch to the tractor draw bar, not to other parts of the tractor, and make sure you use a tow line big enough to handle the job. If you are using a fabric strap as part of the tow line, it should be connected so that if any type of failure occurs while the strap is under tension, the stored energy in the strap will not cause objects to be thrown or propelled toward workers. For example, if you are using a combination of chain and strap, attach the strap to the pulled object and attach the chain to the tractor. A cable that is

Inspect towing cables before use. Cables with damaged areas or broken strands are more likely to fail. Broken wire ends are also more likely to catch on skin and clothing.

Small irrigation pipes that can be moved by hand are often used around farmyards to water lawns and beds.

Small rodents are fond of hiding in irrigation pipes. Check for overhead electrical wires before lifting the pipe to a vertical position to scare them out.

CASE STUDIES: CONTACT WITH OVERHEAD WIRES

1) Two workers at a California farming company were electrocuted when the aluminum irrigation pipe they were lifting made contact with an overhead electrical line. The victims, ages 23 and 53, were found lying on the ground, and a piece of irrigation pipe was lying on top of one victim's chest. There were no witnesses to the incident. Coworkers and the supervisor believe the employees were attempting to remove a rabbit or some other small animal from the pipe when contact was made with the overhead power line.
—from California FACE Report # 94CA001

2) Workers on a large farm were in the process of moving a grain auger when it made contact with a 7,200-volt power line. Two workers (30 and 39 years old) were electrocuted and three other workers were injured.

    The auger can be raised or lowered by a hand crank that is attached to a steel cable pulley system. Common practice for moving the auger is to lower it for stability. The auger was to be moved from the grain-drying bin, approximately 30 feet high, to a different location. To move the auger it had to be raised to allow the top flap to clear the bin and to allow the bottom out of the ground-loading ditch. It is estimated that the auger was raised to a height of 35 feet to enable the workers to back it away from the dryer bin. The five workmen were positioned around the rear of the auger to move it to the new location.

    This incident is almost identical to Georgia FACE Report #86-6 where three workers were electrocuted.
—from Georgia FACE Report #86-07

kinked or frayed is in danger of snapping at that point. Cables that break under tension flail around unpredictably and can cut anything in their path.

Take up slack slowly and exert a steady pull. Ask the operator of the stuck vehicle to drive the same direction while you pull. Taking a running start to try to jerk the load free can stress any tow line to the breaking point. When you disconnect the tow line, make sure all tension is off and that all vehicles or towed objects cannot roll or slide into contact with the person unhooking the tow line.

OVERHEAD ELECTRICAL WIRES

The electrical service lines in many rural areas and farmyards are strung overhead from pole to pole, rather than underground as in newer urban areas. Normally, the lines are so out of the way that they tend not to be noticed. This leads to the risk of hazardous contact when moving equipment or long pieces of metal, such as aluminum extension ladders or irrigation pipes.

PRECAUTIONS AROUND OVERHEAD WIRES

Watch carefully for overhead wires when moving any tall equipment, such as a grain auger, conveyor, fold-up implement, or a tractor with the front-end loader raised.

Check for overhead wires before raising long pieces of metal, such as aluminum extension ladders or irrigation pipes.

Before attempting to repair electric lines broken by farm equipment, move the transfer switch securely to the OFF position.

LADDERS

Ladders are already one of the biggest safety hazards encountered by most people in their day-to-day work and home life, and the hazards are just as common on the farm. At least 300 people a year in the United States die in falls from ladders, and ladder-related accidents account for about 100,000 injuries per year. A long fall isn't necessary for major injury. Statistics indicate that a fall of 11 feet or more will kill 50 percent of victims.

Make sure that access ladders are securely placed to prevent both falls and the risk of being stranded atop a building if the ladder falls down. This ladder is chained to the roof catwalk.

Due to long-term exposure to the elements, the main breaker switch on older transfer boxes may no longer operate smoothly. Before you start to repair broken wires, confirm that the switch is fully in the OFF position. Tingling in your hands or arms when you grab the wires is not a good test—your heart is directly in the path of the current.

A safety cage around ladders helps reduce risk, especially around access ladders that are frequently used.

YARD AND WORKSHOP

With basic practical precautions, you can avoid many hazards in a typical farmyard and workshop. For example, while compressed air is vital for many shop tasks, using high-pressure air for cleaning dust off your clothes is a very dangerous practice. Compressed air that enters the bloodstream through a break in the skin or through a body opening can cause an air bubble to form in the bloodstream (embolism), which can lead to coma, paralysis, or death. Horseplay with the air hose has been a cause of some serious workplace accidents.

Above: *Install handrails on all elevated access platforms to prevent falls from accidental steps backward or slips while turning.*

Left: *Where irrigation valves are electrically controlled, keep both electrical and water connections in good repair with no leakage. You don't want to be near a site where water and electricity mix!*

To keep children from scaling buildings with permanent access ladders, keep the end of the ladder above a child's reach.

Install solid steps with handrails, especially where you often carry feed pails, bales, or heavy items.

Putting simple barriers around hazardous sites helps prevent hazards that could occur when accidentally backing into or running over pipes with trucks and implements.

Careless removal of a radiator pressure cap can blow steam and coolant into your face and eyes. Let the engine cool down completely and turn the cap slowly to release pressure.

Welding acetylene bottles must always be stored and used in an upright position to prevent unexpected fires.

Above: *Transporting acetylene bottles in a horizontal position may allow liquid acetone to escape around the valve. A fire containing liquid acetone spreads quickly and is very hard to extinguish.*

Left: *In the workshop, do not rely on jacks alone to support a weight. Use solid stands to block the weight from falling.*

When you're using your ATV around the farm, leave the stunts for those times when you have a proper helmet and riding gear.

CHAPTER 2
ROLLOVERS AND TURNOVERS

▶ *Tractor rollovers are the single deadliest type of injury incident on farms.*

▶ *One of every 10 tractor operators overturns a tractor in his or her lifetime.*

▶ *Runover accidents caused by bypass starting or falling off a moving tractor cause the greatest number of tractor-related casualties.*

The application of engine power to various farm tasks, such as tillage, planting, crop cutting, and fencing, makes the work tremendously less laborious. Without tractor power and purpose-built implements, it's unlikely that anyone would look forward to work around the modern farm, whether it is a small-scale or lifestyle farm or a large commercial farm.

But the ease with which tractors and implements can accomplish heavy work can also make it easy to be careless around them and forget about how important it is to think about safety when using them. Danger is not necessarily decreased with smaller tractors and implements. Even a maneuverable, handy little compact tractor weighs several thousand pounds. If it tips over and lands on you, it can kill you just as effectively as a more intimidating 350-horsepower, 20-ton, four-wheel-drive field tractor.

Tractor-related accidents result in more serious, disabling injuries and fatalities than any other type of agricultural safety situation. Tractor accidents are linked to more than half of all farm-related deaths. In the United States, there are approximately 250 tractor rollover fatalities per year, according to NIOSH data.

Rollover incidents are particularly deadly due to serious crush injuries to the head, chest, and pelvis when operators are pinned under the tractor. Incidents may occur far from sources of help and the injured operator may not be discovered for many hours. Similar data from Canada illustrates the deadliness of the problem. Rollover injuries involving farm tractors

account for 22.8 percent of all agricultural work-related fatalities, but only 2.3 percent of hospitalized injuries, because most victims are killed at the scene.

Accidents where the victim is run over by the tractor, the implement, or both, are just as deadly and even more common. These accidents typically occur when the victim is run over while starting the engine from the ground and standing in front of a wheel or when falling off a moving tractor. In a study of 32 fatalities involving tractors from January 1993 to July 1998, 56 percent of the deaths involved the victim being run over.

A LOOK AT TRACTOR STABILITY

To be useful around the farm, tractors need:
- Tremendous torque to be able to pull heavy loads from a standstill.
- Lots of weight (as a rule of thumb, about 100 pounds per horsepower) for the tires to grip the soil enough to develop their best pulling power.
- Plenty of ground clearance so they can be operated in crops and over rough ground.
- A relatively short wheelbase in order to be as maneuverable as possible.
- Easy steering to prevent operator fatigue.
- Enough road-speed capability to be able to move around the yard or between fields without taking excessive time.

The combination of these necessary design features logically creates some important consequences in terms of stability. The massive torque needed to pull

Rollover protection structures (ROPS) can protect the operator in a tractor rollover. A seatbelt keeps the operator from being thrown from the protective zone. If an unbelted operator is thrown out (or tries to jump out) during a rollover, he or she can be crushed by the t ractor or an implement. Pennsylvania State University Agricultural & Biological Engineering Department

Some of the same design factors needed to make tractors useful in normal situations make them dangerous if they are not operated with due care or outside normal parameters. AGCO Corporation

A steel-tracked tractor carries a lot of its weight low, which makes it possible to work on slopes much steeper than where a farm tractor could safely operate. But in spite of their greater stability and tremendous pulling power, the slow travel speeds and damage inflicted on roads and yards makes steel-tracked crawlers unsuitable for most farms. The very hilly Palouse region of Washington is one notable exception.

Skid-steer loaders also carry their weight low, which improves their side-to-side stability. But they also have very little ground clearance, which makes them unsuitable for pulling implements. The very short wheelbase helps make them super maneuverable, but they are also prone to frontward or backward tipping.

heavy loads at the drawbar can lead to the tractor flipping backward quickly if the load is improperly hitched, such as to a point higher than the drawbar. A lot of weight carried relatively high above the ground because of the need for ground clearance means the tractor has a high center of gravity. Naturally, this makes the tractor more susceptible to tipping on sloping ground, such as ditch banks. When a tractor does tip, its weight will make pudding out of anyone caught beneath. The short wheelbase that helps make the tractor maneuverable at low working speeds also makes the steering remarkably quick at road speeds. On the road, a small movement of the wheel can lead to a surprisingly sudden movement of the tractor toward the ditch.

SEVEN DANGER ZONES

Knowing the above characteristics of tractors helps you to understand why particular types of terrain and tractor operating situations are noted for particular types of tipping accidents.

1. Side slopes
• Tall grass can hide potholes and large rocks, so walk the slope first to check for hazards.

CASE STUDY: TRICYCLE-TYPE TRACTOR INSTABILITY

A 40-year-old tricycle tractor not equipped with a ROPS or a seatbelt had a bush-hog attachment. On the morning of the incident, the operator and his brother were mowing a portion of a pasture. At about 9:30 a.m., the victim started mowing at the bottom of a steep slope and progressed toward the top of the ridge. In order to make turns he had to circle around three dogwood trees at the top of the ridge. The victim was pinned underneath a rear tractor tire after it overturned.

—from Kentucky FACE Report #95KY04701

A narrow front-wheel configuration made tractors very useful in row-crop farming tasks such as cultivating corn and soybeans. But this design also increases tipping hazards, so equipping this type of tractor with a front end-loader is not recommended.

- Be sure the tractor's brake pedals are locked together and test the brakes before going on to the slope. If only the uphill-side brake is applied on a side slope, the tractor will slew upward and increase the risk of a sideways rollover.
- Wet grass or loose soil on the slope can lead to the tractor sliding sideways, then tipping sideways if the tires stop sliding.
- Driving the tractor near the top of a slope may put the tractor on less stable soil. If the soil gives way, the tractor can start sliding down the slope and tip over sideways.

- If it's practical, work up and down the slope rather than across it. Travel down the slope going forward; travel up the slope in reverse.
- When traveling across a side slope, keep the steering wheel steady while looking back at the implement. If you accidentally turn up the slope while looking backward, this could cause a sideways rollover.
- If work on side slopes is continually necessary, set the tractor wheels as wide as possible. For increased stability, consider using a low-profile tractor (e.g., orchard special) or installing smaller-diameter wheels on an existing tractor to lower the center of gravity.

Another type of single-front-wheel (tricycle) tractor is not recommended for use with a front-end loader. Exercise extra caution when operating this design on anything other than smooth row-crop ground.

2. Climbing a slope

- Climb with the heavy end up the slope (in reverse).
- Wet grass or loose soil on the slope can lead to the tractor spinning the rear wheels, then flipping backward if the tires suddenly find good traction.
- Up-slopes often have a very steep (sometimes near vertical) final few inches at the top. This sudden increase in slope increases the tipping hazard.
- Be sure the tractor's brake pedals are locked together and test the brakes before climbing the slope. If only one brake is applied on a slope, the tractor will slew to one side and increase the risk of a sideways rollover.
- If the tractor starts rolling backward, avoid sudden application of the brakes. Tractor brakes work only on the rear axle, and panic braking with the nose tilted up dramatically increases the risk of a backward rollover.
- If you can only get out of the field and on to the road by traveling up a steep ditch bank, consider either building a less-steep ramp or traveling to and from the field in the ditch bottom and avoid slopes altogether.
- If the tractor has an implement mounted on the three-point hitch, it may strike the ground as the tractor's front wheels rise up the slope. This may leave the rear wheels spinning freely and unable to

drive the tractor forward; you may need a tow out of the ditch. Raise the implement to full height before starting up the slope, then start up slowly so you can be sure the implement won't get hung up. Once the tractor's rear wheels are on the slope, you can lower the implement slightly to act as a stabilizer that prevents the tractor-implement combination from tipping backward as easily.

Rear tread width has been made easier with various adjusting methods, such as the rack-and-pinion method on many John Deere tractors or the power-adjusted variable tread (PAVT) system shown here. The manual for your tractor explains how to adjust tread width.

In situations where tipping may be a concern, such as working on a slope, side-to-side stability can be improved by setting the wheels as wide as possible. Wheels may be able to move farther apart on the axle, or in some cases they can be reversed (dished out) to achieve a wider setting.

Front axles on many tractors have some provision for adjusting width. To improve stability, use the maximum width practical for your job.

Above and left: *Many tractor models are also available in specialty configurations that have low-profile wheels. This configuration increases stability, and maximizes clearance in tight work, such as between orchard trees.* AGCO Corporation

3. Descending a slope

- Walk the path first to check for potholes or large rocks that may be hidden by tall grass.
- Descend with the heavy end up the slope (i.e., travel forward).
- Travel straight down the slope, not at an angle. Descending at an angle introduces the problem of traveling across the slope and leads to an increased risk of a sideways roll.

- Be sure the tractor's brake pedals are locked together and test the brakes before descending. If only one brake is applied on a slope, the tractor will slew to one side and increase the risk of a sideways rollover.
- If traveling with a load in the tractor's front-end loader, keep the bucket close to the ground. Be ready to raise the load as the bucket reaches the bottom of the slope to avoid jamming the bucket into the ground.

Left and below: *Here are two versions of the popular Ford 8N/9N vintage tractor. The smaller-diameter wheels on the less common low-profile model (blue) bring it lower to the ground and increase its stability. Dual rear wheels also improve stability.*

- If you can only get off the road and out of the field by traveling down a steep ditch bank, consider either building a less-steep ramp or traveling to and from the field in the ditch bottom and avoid slopes altogether.
- If traveling with an implement mounted on the tractor's three-point hitch, raise the implement as high as possible. Otherwise, when you reach the bottom of the slope the dragging implement may jack the tractor's rear wheels off the ground and cause a loss of traction. It's not a rollover danger, but you may be thrown forward and need a tow to get the tractor rolling again.

4. Sharp turns

- Before traveling above field speeds (3.5 to 4 miles per hour) be sure the tractor's brake pedals are locked together and test the brakes to make sure both are being applied evenly. If only one is applied at speed, the tractor will slew to one side and increase the risk of a sideways rollover.

Unless you are turning the tractor on relatively flat ground and at the low speeds used for field work, keep the brake pedals locked together.

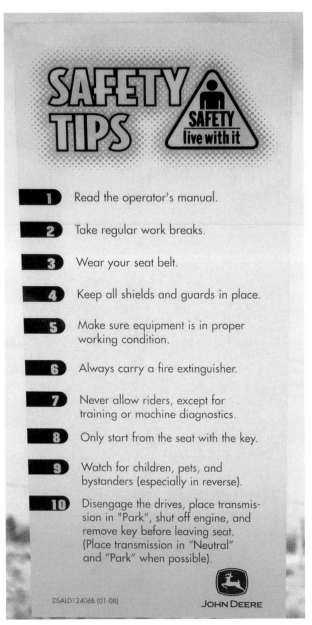

SAFETY TIPS — SAFETY live with it

1 Read the operator's manual.

2 Take regular work breaks.

3 Wear your seat belt.

4 Keep all shields and guards in place.

5 Make sure equipment is in proper working condition.

6 Always carry a fire extinguisher.

7 Never allow riders, except for training or machine diagnostics.

8 Only start from the seat with the key.

9 Watch for children, pets, and bystanders (especially in reverse).

10 Disengage the drives, place transmission in "Park", shut off engine, and remove key before leaving seat. (Place transmission in "Neutral" and "Park" when possible).

DSALD12406B (01-08)   JOHN DEERE

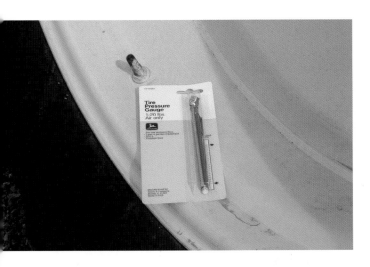

Maintaining correct tire pressure is critical to maximizing operating efficiency and safety. Since tractor tires typically run at low pressure, a low-pressure gauge ensures more dependable results.

This sticker helps put safe operating methods front and center while you are on the tractor. It can be ordered at no charge from John Deere dealers and is part number DSALD10902.

Traversing a slope can cause a tractor to tip sideways. Factors that suddenly increase the danger of tipping include a downhill wheel falling into a pothole or animal burrow, an uphill wheel jarred upward by a stone or dirt pile, or a lifted implement that raises the overall center of gravity.

Climbing a slope in a forward direction can cause the tractor to tip backward. Factors that increase the danger of tipping include the uphill axle getting jarred upward by rough ground or the operator applying the brakes if the tractor starts rolling backward.

It is generally harder to tip over a tractor forward when descending a slope because the center of gravity for most tractors is well toward the rear axle. However, certain conditions can make a forward tumble possible.

The high center of gravity on tractors means a sharp turn at elevated speeds can cause a sideways rollover. A load carried high in the front-end loader raises the center of gravity and increases the tendency to tip while making a turn.

- Avoid sharp turns at road speed or on slopes.
- Watch out for ruts or potholes that could jerk the steering wheel out of your hands and lead to an unexpected sharp turn. Either avoid the ruts and potholes or slow down so you can retain control of the steering.

5. Starting off with heavy loads
- Hitch heavy loads only to the drawbar.
- Rollover protection structures (ROPS) should not be used as a hitching point for a chain, rope, or cable. Pulling with the ROPS could damage it and result in a rear overturn.
- Take up any slack in load connection slowly and smoothly.
- Stay ready to cut power immediately if the front end of the tractor starts to rise.
- Do not pull at an angle because it adds the risk of a sideways rollover.

6. Road speeds
- Pick a safe travel path and stick to it. If you have to pull over to let traffic by, look for a safe pullout or wider spot in the road and slow down or stop.
- If the road you travel has lots of high-speed traffic, consider traveling to and from the field in the ditch bottom or ends of fields and avoid roads altogether.

- Before traveling at road speeds, be sure the tractor's brake pedals are locked together and test the brakes to make sure both are being applied evenly. If only one brake is applied at speed, the tractor will slew to one side and increase the risk of a sideways rollover.
- Avoid sharp turns at road speed.
- Large bumps (e.g., at railway crossings) can cause the tractor to jolt so severely that steering control is momentarily lost. Slow down for big bumps in the road.
- The slow road speed (about 15 to 20 miles per hour) of most tractors can quickly lead to inattention to the road. Sometimes farmers are known to get so absorbed in checking the crops beside the road that they veer off the road and into the ditch.

7. Edges of waterways
- Test the ground by walking it first. Soil close to the edge of a natural or man-made waterway is likely to be crumbly and have reduced load-bearing capability. If the soil suddenly gives way, this increases the risk of the tractor tipping down the bank.
- Avoid sharp turns near the edges so a tire doesn't suddenly slip over the edge of the bank.
- Keep the steering wheel steady when you turn to look back at the implement. If you accidentally turn toward the bank while looking backward, it could initiate a sideways rollover down the bank.

If turning with a load in the front-end loader, keep the load low—just high enough to clear the ground. Turning with the bucket raised high increases the risk of sideways rollover.
AGCO Corporation

- Work at a reduced speed so that any problems that develop happen at a slower pace and give you a little more time to take corrective action.
- If the implement projects to one side, make the initial pass with the implement projecting toward the watercourse. This positions the tractor away from the edge.

ROLLOVER PROTECTION STRUCTURES

Rollover protection structures (ROPS) are roll bars or roll cages designed for farm tractors. The goal of the ROPS is to absorb the impact energy without excessive deformation to create a zone of protection for the operator. However, to work effectively the operator must stay within the protective zone. If the tractor rolls over and the operator falls out of the seat or tries to jump clear, the ROPS cannot do the job for which it was designed.

When used with a seatbelt, the ROPS will prevent the operator from being thrown out of the protective zone and crushed from an overturning tractor or the equipment mounted or hitched to the tractor. A ROPS also protects the tractor by often limiting the degree of rollover.

The use of ROPS and a seatbelt is estimated to be 99 percent effective in preventing death or serious injury in the event of a tractor rollover. If you're interested in leaving a legacy, consider this: the one-time

Hitch heavy loads only to the drawbar. This achieves optimum pulling power by proper transfer of the load to the tractor chassis, and it decreases the risk of backward rollover.

Do not hitch anything to other parts of the tractor, especially above the height of the drawbar. A backward rollover from improper hitching tends to happen much faster than victims can react. Research shows that it generally takes only about 3/4 of a second to reach the critical tipping point and about another 3/4 of a second for the tractor to crash backward.

CASE STUDY: OVERTURNS RELATED TO PULLING LOADS

1) A 33-year-old male farm worker died as a result of multiple head and torso injuries sustained during a rear rollover of the 1958 tractor he was using to pull a pickup truck filled with wood. The tow chain had been hitched high on the back of the tractor.

2) A 42-year-old female farmer died from chest trauma when a 1970 tractor she was using to pull a loaded pickup truck out of the snow overturned to the rear. The tow chain had been attached at the top link connection of the tractor's three-point hitch.

3) A 13-year-old boy sustained fatal massive head trauma when the 1953 tractor he was using overturned to the rear while pulling a felled 18-inch-diameter tree that was still partially attached at the stump. The tow chain had been hooked directly around the rear axle.

—from "Fatalities Associated with Improper Hitching to Farm Tractors —New York [State], 1991-1995."

When road traffic stacks up behind a tractor, the operator tends to move as far to the right as possible to let vehicles pass. But if the tractor and/or implement wheels on one side run off the road and into softer gravel or dirt, the increased rolling resistance on that side can jerk the tractor toward the ditch. The result is a sudden, dangerous plunge down and across a slope.

effort in installing a ROPS will, for the life of the tractor, protect whoever drives it—perhaps your own child or grandchild.

Three types of ROPS frames are available: a two-post frame (with solid fold-down versions), a four-post frame, and a ROPS with enclosed cab. They all serve the same function and protect the operator in case of a tractor rollover.

Above: *Working too close to the edge of a stream, river, canal, or other waterway increases the risk of a sideways rollover and adds the danger of drowning in the water or mud at the bottom of the slope.*

Below: *As a rule of thumb, stay at least one tractor-width away from the edge. If the tractor tips, it has a better chance of landing on flat ground, not on the slope.*

CASE STUDY: TRACTORS AT ROAD SPEEDS

A 16-year-old farm worker died of injuries when the tractor he was driving overturned and crushed him. The victim was towing a small hay baler on a two-lane public paved road. Just prior to the incident, a motorist observed the victim driving the tractor and baler with the right wheels of both partially on the road's gravel shoulder. The tractor then turned right toward the road's steeply banked shoulder. As the tractor began to overturn down the bank, the victim attempted to jump from the tractor in the same direction as the overturn. The motorist who witnessed the incident found the victim pinned under the tractor. Emergency medical technicians were called and responded to the incident. The victim died at the scene.
—from Washington FACE Report # 99WA05601

Between 1967 and 1985, U. S. farm-tractor manufacturers provided ROPS as optional equipment on most tractor models. This meant that anyone looking for a new tractor had to add the cost of a ROPS on to the base price of a tractor. Because most farmers are very cost-conscious, few added ROPS as an option. Even fewer pre-1967 tractors have ROPS, yet many of these tractors are still in use. Beginning in 1986, American tractor manufacturers began voluntarily adding ROPS on all farm tractors over 20 horsepower sold in the United States.

Holes should never be drilled into the ROPS frame, nor should a piece of steel be welded onto the frame, because this will weaken the frame in unpredictable ways. If lighting or other attachments are needed, they can be clamped onto the ROPS.

Many farm tractors were built before ROPS was commonly used, and the percentage of tractors in use without ROPS is high because farm tractors are often in use for 30 to 40 or more years.

Foldable ROPS, such as the tractor at left, are now available to solve the problem of fitting the tractor into smaller storage spaces, such as a garage with a low doorway.

If a tractor with ROPS does overturn, the ROPS should be replaced because it is specifically designed to bend to absorb the energy generated by the tractor contacting the ground. ROPS are only designed and certified to withstand a single overturn.

Many farm tractors manufactured after 1967 can be retrofitted with a ROPS. Tractor companies and aftermarket manufacturers have designed and developed ROPS for most tractor models. Manufacturers and tractor supply houses offer low-cost retrofit ROPS kits for tractors manufactured from the mid-1960s to 1985. ROPS for many older and smaller tractors can be purchased for $600 or less. Agricultural equipment dealers are approved to install a retrofit ROPS and seatbelt. Installation charges are normally in addition to the cost of the ROPS.

A listing of ROPS retrofits for farm tractors manufactured since 1967 has been compiled by the National Farm Medicine Center at Marshfield, Wisconsin, in a publication called *A Guide to Agricultural Tractor Rollover Protection*. This guidebook is available on the web at http://research.marshfieldclinic.org/nfmc/resources/rops/tractor.asp. Local equipment dealers should also have information on ROPS retrofitting for their brands of tractors.

A seatbelt is an integral part of the ROPS because it keeps the operator within the protective zone created by the roll bar or roll cage. A ROPS alone will not provide full protection to the operator when there is a tractor overturn. Without a seatbelt the operator will not be confined to the protective zone and may be crushed by the tractor or the ROPS itself.

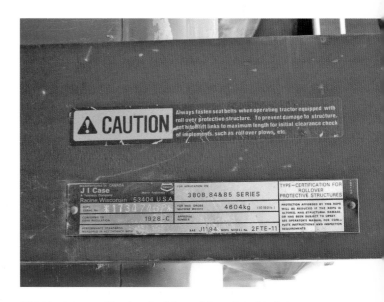

When buying a used tractor, check for a plate such as this that indicates a tested ROPS design. Homemade ROPS are generally not recommended due to lack of certainty on how the finished product will perform.

RUNOVERS

- Runovers are second only to tractor rollovers as a cause of death on farms.
- Being run over by the tractor or attached equipment accounts for about 16 percent of all agricultural fatalities and 5 percent of hospitalizations.
- Among adults, bypass starting (operator standing on the ground during the start) is the leading cause of tractor runover accidents.

On tractors with bad wiring, broken starter switches, weak batteries or starters, and/or missing or bypassed safety interlocks, many operators are tempted to work around these difficulties by jump-starting or short-circuiting part of the starter wiring.

Clear information about the dangers of bypass starting can be put where operators may be tempted to make unsafe engine starts. This sticker can be ordered at no charge from John Deere dealers and is part number DKLD285.

CASE STUDY: ROPS AND SEATBELTS

A 60-year-old male minister and part-time farmer began mowing a 15-acre field near his home at about 11:00 a.m. He was operating a farm tractor with a side-sickle bar mower attached to the right side and proceeded around the perimeter of the field. One edge of the field dropped off steeply to a ravine. While making the second pass around the field the ground under the left wheels gave way and caused the tractor to roll over. The tractor came to rest on the victim and caused massive internal injuries. A Roll Over Protection Structure (ROPS) had been retrofitted to the 1962 tractor about 10 years prior to the incident. The tractor did not have a seatbelt. The FACE investigator concluded that to prevent future similar occurrences, tractor operators should wear seatbelts when operating ROPS-equipped tractors.
—from Kentucky FACE Report # 94KY09101

Tractors usually have various safety devices built into the starting system to prevent unsafe startups. The transmission and PTO may have to be in neutral or the clutch disengaged before the starter is energized.

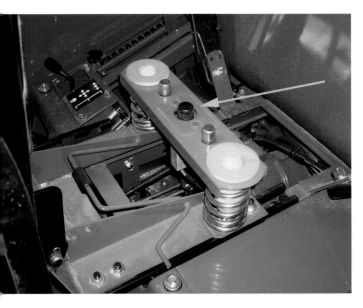

Newer tractors have a switch that only allows starting with the operator's weight on the seat. Do not disable the switch if you find that it stops the tractor when you stand up or lean over to look at the implement. Keep the safety benefits intact by pressing the backs of your legs against the seat to hold it down.

Runovers when starting from the ground can be prevented with an inexpensive remote starter switch. The switch wires are clipped to the starter solenoid terminals.

CASE STUDY: GROUND STARTING

1) A 60-year-old farmer was killed when he started his tractor while standing on the ground to the left of the tractor or while standing on the side step of the tractor. There were no witnesses to the event, but it appears that the farmer was standing on the ground, holding in the clutch with his left hand, starting the tractor with his other hand, then released the clutch and assumed the transmission was in neutral. The tractor originally had a safety switch built into the transmission, but it had been bypassed by the previous owner of the tractor. The farmer was found trapped partially under the left-rear wheel of the tractor and dead at the scene.
—from Iowa FACE Report #98IA073

2) A 51-year-old farm worker died from injuries sustained when he was run over by the farm tractor he was operating. Because a key had previously broken off in the ignition switch and the switch had not been repaired, to start the tractor the driver used a screwdriver to create an electrical short between the tractor's battery and starter. This was an operation that required the worker to position himself directly in front of the inner-right front wheel while reaching into the engine compartment. The tractor lurched forward, ran over the driver, knocked down a bystander, and veered toward a roadway before it was stopped by another person. The medical examiner listed the immediate cause of death as multiple head and trunk injuries.
—from Oklahoma FACE Report #01-OK014-01

With the remote starter switch connected, the operator can easily step away from the runover danger area. A simple press of the switch cranks the engine.

CASE STUDY: HELPER HAZARDS WHILE HITCHING

The 60-year-old wife of a farmer in Colorado was assisting her husband, who was doing farm chores for a neighbor. The brakes on the tractor were defective, which was known by the operator. The operator had instructed his wife to guide him back to the feed grinder and insert the draw pin. As the tractor proceeded backward with a jerking motion, the victim's foot either slipped or was caught under the rear tire. As she fell down, the tractor traveled over the length of her body. The coroner listed cause of death as massive traumatic abdominal and chest injuries.

—from Colorado FACE Report #90CO0058

The problem is that starting the tractor this way usually means the operator is standing right in front of a heavy driving wheel. If the tractor is inadvertently left in gear, the tractor can suddenly run over the operator or bystanders when the engine catches. Evidence from accident investigations shows there is almost no time for the operator to react in such situations.

SAFETY PRECAUTIONS WHEN STARTING OR BOOSTING THE TRACTOR

- Disengage the clutch on manual-shift tractors.
- Take the transmission out of gear or place the range lever in PARK.
- If booster cables are used, connect them to the dead battery and not directly to the starter.
- Start the engine from the operator's seat with the starter, or use a remote starter switch.
- If the starter circuit is broken or inoperable, repair the fault. Do not use short-circuit methods. For information on how to diagnose and repair starter circuits, please refer to the book *How to Keep Your Tractor Running*.
- For women, runover accidents often occur as they assist their spouses in hitching equipment to tractors.

If any temporarily inoperative part of the tractor could result in a safety hazard, leave information where other operators can see it, or simply as a personal reminder. A tag can be ordered at no charge from John Deere dealers and is part number DKCD265.

CRYSTAL ZIMMER

When Crystal Zimmer was 2 1/2 years old, she ran out from the side of her house into the pathway of the riding lawnmower her mother was operating. She lost her right heel and fifth toe, half her left heel, and suffered severe lacerations to both calves. Surgery and rehabilitation enabled her to walk again, but the incident will have an effect on her for a lifetime.

"I have my limits. I have to know when to stop so I don't overexert myself. They had to put pins in my right ankle, and if I overdo it, I'll have pain. Because my feet are misshapen, I have to wear molded insoles in my runners, and I can't wear heels. I need shoes with great support that lace up tight around the ankles so my feet don't shift. This prevents blisters and sores forming on the bottoms of my feet. I often get stares—whether they are curious or rude or derogatory comments—but for the most part, people are very understanding."

A helper standing behind a running tractor is in a very vulnerable position if the tractor unexpectedly moves. Since the operator is usually facing backward in this situation, it is too easy for a foot to slip off the clutch or for the operator to make an incorrect movement of the controls.

Inadvertent movement of the tractor can easily injure a person in the close quarters between the tractor and implement. Women are at a higher risk than men of being run over by tractors and other farm machinery, according to data from the U.S. National Safety Council.

SAFETY PRECAUTIONS WHEN A HELPER IS BEHIND THE TRACTOR

- Use a low throttle setting to minimize sudden movements.
- For tractors with cabs, open the rear window to permit clear, instant communication between you and the helper.
- While you are moving in reverse, have the helper stand off to the side.
- Back up slightly more than necessary, then shift to a forward gear before the helper moves in between the tractor and implement.
- Have the helper insert hitch pins while you creep forward. If the tractor is in forward gear, it cannot back over the helper if your foot slips off the clutch or creeper pedal.
- If the tractor hitch points need to be moved while you are moving backward, have the helper use a long stick or bar so they can remain clear of the tractor.
- Pull forward or raise a three-point hitch implement only after the helper has moved away from the tractor and implement.

CASE STUDY: RUN-OVER INJURIES TO CHILDREN AND YOUTHS

1) A one-year-old girl in Kentucky was run over by a farm tractor driven by her father, who was spreading mulch around trees lining a farm road. He drove the tractor along the road and stopped every few feet to apply mulch. In the late afternoon he took a break with his wife and three children, who came to visit. As he prepared to resume work, his wife and children walked to a nearby creek. He saw his wife and two of the children, and assuming the third child was also with his wife, engaged the tractor. His daughter was run over by the right-rear tractor tire and died instantly from blunt impact to the head, trunk, and extremities and crushing head injuries.

2) A two-year-old girl in Iowa was killed on the family hog farm when she was run over by a tractor driven by her father. As the father was loading hogs into a livestock trailer attached to the tractor, his wife was assisting and the child was playing nearby. When he drove the tractor forward, the right-front wheel ran over the child's lower torso. The child remained conscious and crying after the incident and was airlifted to a regional children's hospital where she died four hours later from internal bleeding.

—from "Deaths Among Children Aged Less Than or Equal to 5 Years from Farm Machinery Run-overs—Iowa, Kentucky, and Wisconsin, 1995–1998, and United States, 1990–1995," Centers for Disease Control and Prevention

Among youths and children, runover accidents result from other causes: a child bystander is run over by farm machinery, or a child who is an extra rider on a tractor falls from the tractor and is run over. These two causes account for 75 percent of all fatalities in young children on the farm, according to Canadian Agriculture Injury Surveillance Program data. These tragedies are worse because it is generally a parent or relative who is operating the tractor that runs over the child.

SAFETY PRECAUTIONS AROUND NONOPERATORS

- No riders unless your tractor is equipped with an approved trainer seat inside the ROPS safety zone.
- No riders in the bucket of the front-end loader, the platform, steps, fenders, hood, or standing on the drawbar.
- Before moving the tractor, visually and audibly confirm the location of anyone nearby and don't move until they are out of your path.

CHAPTER 3
FRONT-END LOADERS

▶ *Of 22 patients admitted to the Plains Health Center in Regina, Saskatchewan, from January 1979 to April 1986 with spinal injuries due to farming accidents, seven had injuries related to tractor-mounted, front-end bale loaders.*

A front-end loader (FEL) is often one of the first optional attachments installed on a farm tractor and makes the tractor capable of lifting and transporting objects and material around the farm. However, loads in the bucket always affect the stability characteristics of the tractor. For safe operation and reduced stress on the loader components, the operator should be aware of the changes and how to minimize their effects.

HOW BUCKET LOAD AFFECTS STABILITY

Viewed from the side, a tractor with a front-end loader behaves like a lever and fulcrum system. At one end of the lever is the load in the bucket, while the weight of the tractor is the load at the other end. The fulcrum, or pivot point, is the front axle of the tractor. That is why a load in the bucket tends to lift the rear wheels. If the

The heavier the load in the bucket, the greater the effect on tractor stability.

A 51-year-old Iowa farm wife was killed while helping her husband and son build a new home (for the son) on their farm. The woman and her 21-year-old son were standing next to the exterior foundation of the home and covering drainage tubing with gravel. The woman was holding down the tile with her feet while her son was shoveling gravel out of a partially filled front-end-loader bucket on a tractor his father was driving. The edge of the bucket was approximately 3 feet from the foundation wall when this work began. The father had the tractor in gear with the clutch pushed in and his foot on the brakes, sitting on a slope of about 18 degrees downward toward the house foundation. When he began to raise the bucket, the farmer heard his son yell at him and realized the tractor was creeping forward toward the foundation despite his pressure on the brakes. He saw his son being pinned to the wall with his back toward the tractor, then realized that his wife was also being pinned to the wall with the other corner of the bucket. He immediately put the tractor in reverse and backed up, at which time both his wife and son fell to the ground. The son received only minor injuries to his back, but the farmer's wife was killed instantly from internal vascular rupture.

—from Iowa FACE Report #96IA061

tractor is moving, rear-end bounciness also increases if the tractor goes down a slope, or if the wheels hit a bump or hole.

One crucial consequence of this levering action is that lightening the load on the rear wheels of the tractor also reduces the effectiveness of the brakes. Farm tractors only have brakes on the rear wheels. If a heavy load raises the rear wheels, no braking power is available. The tractor may roll freely forward or backward, depending on the slope of the ground.

Viewed from the front or back, the FEL has another type of lever effect. The pivot point is where the tires touch the ground, and the force at the end of the lever is the load in the bucket. The higher the bucket is raised, the longer the lever, so the more tipping force the tractor receives if the load moves away from straight above the middle of the tractor.

In addition, having the front wheels set close together increases the effect of any sideways tipping force, while wheels set farther apart makes the loader-equipped tractor more stable. The stability benefits of wider axle settings are also the reason to avoid using front-end loaders on tricycle-type tractors.

Worn-out loader joints make the loader more erratic in operation and increase the risk of accidental pivot-pin breakage. Keep joints well greased for continued safe, predictable operation.

A tractor with a FEL is subject to both sideways and front-to-back tipping forces. Turning increases the effects, and the faster the travel speed, the more pronounced the effect.

The final stability point of view to be considered is at the bucket itself. If the loader arms are raised very high without adjusting the forward-and-back angle of the bucket, there is increased risk that the load in the bucket may roll back toward the operator. Many accidents of this nature occur when handling large round hay bales in an ordinary FEL.

DANGER ZONES AND HOW TO DEAL WITH THEM

1. Heavy loads

• Consult the manual to find out the recommended maximum bucket load.
• Fill the bucket less when handling very heavy materials, such as damp sand.

Be extra careful if lifting a load or pulling backward with a chain attached to the bucket. If the applied load is a bit off center, the tractor may lurch or tip sideways.

The heavier the load, the more lifting force is applied to the rear of the tractor.

The higher the load in the front-end loader (FEL) bucket is raised, the more sideways leverage is exerted on the tractor if it is driven on even a slight side slope or if a wheel hits a bump or hole. Tipping force can be increased if the load happens to shift sideways in the bucket.

- Add weight to the rear of the tractor with wheel weights or by attaching a ballast box or heavy implement to the three-point hitch.
- Lift the bucket slowly to prevent sudden tractor instability.
- Keep the bucket low when traveling.
- Reduce travel speed on slopes, in turns, and on rough ground.

2. High lifts

- Use high lift only on flat ground.
- Raise the load slowly to prevent sudden tractor instability.
- For loads that can roll or shift, such as hay bales, use only loaders equipped with a grapple that prevents backward rolling.
- Lower the load for traveling.
- As you lift the loader arms, adjust the angle of the bucket so that the floor of the bucket does not tip too far backward. Self-leveling loaders automatically keep the bucket level.
- Watch out for overhead electrical wires and other obstacles when the loader is raised.

continued on page 52

Front wheels that are farther apart help increase stability. If your tractor has adjustable spacing on the front axle, use the widest practical setting when doing loader work.

Larger tractors are often equipped with a system to keep the bucket level while it is being lifted. If your loader is not equipped with the system, practice using a combination of lever movements to trim the bucket level while lifting.

If the rear end of the tractor starts to lift as the loader is raised, or if it feels bouncy as the tractor moves, there is too much weight in the bucket and/or too little weight at the rear of the tractor. Stop lifting, adjust the load, and add ballast to the rear.

Adding weight to the rear of the tractor for loader work can be as simple as attaching a suitably heavy implement to the three-point hitch. If you do this, be careful when backing up or turning because the implement may strike bystanders or objects.

A load raised very high increases instability of both the tractor and the load within the bucket. For example, hay bales may roll backward out of the bucket and injure or kill the operator.

A grapple or similar structure at the rear of the bucket helps prevent the load from spilling backward toward the operator as the bucket is lifted.

A manure bucket can be attached to the loader for easier, more efficient, and safer handling of loose material.

Remove the bucket when the tractor is used for field work, but keep the loader arms attached. This improves tractor weight balance, increases visibility for the operator, prevents interference with headlights, and improves safety.

CASE STUDY: BUCKET INTERFERENCE WITH FIELD WORK

A 62-year-old farmer died when she was thrown from the farm tractor equipped with a front-end loader. There were no eyewitnesses, but physical evidence indicated the victim raised the plow and lowered the front-end loader bucket into the ground while the tractor was still traveling in a forward direction. The bucket struck the soil with force and penetrated the soil to a sufficient depth to completely stop the forward motion of the tractor. The force of the dual-drive wheels and traction tires spinning continuously against the resistance of the front-end loader anchored in the ground caused the tractor to buck violently. The victim fell from the tractor seat and apparently struck her head as she fell.

—from Oklahoma FACE Report #01OK061

Newer tractors are often equipped with a lockout to prevent inadvertent movement of the loader arms. However, this lock will not block loader movement due to leaking rams.

Continued from page 47

3. Field work

• Remove the loader if it is feasible. Many loaders have a "quick detach" design that allows easy removal and attachment without tools.

• If the loader is not removed, remove the bucket.

• If possible, lock out the loader controls.

4. Riders in the bucket

• Although it looks like a convenient place to ride, especially for children, the front-end loader is not a suitable people mover. Inadvertent movement of the loader controls could put riders in danger of having their legs drag on the ground, especially at road speed. But an even worse danger is that the rider could be bounced out of the bucket when the tractor hits a bump in the road.

5. Working around front-end loaders

• When working on or around front-end loaders, block the loader with support stands or solid blocks of wood before putting any part of your body in a position where it could be crushed by the loader arms. Especially in older tractors with worn hydraulic seals and pumps, hydraulic leakage may cause the loader arms or bucket to creep unexpectedly.

CASE STUDY: UNEXPECTED MOVEMENT OF LOADER ARMS

The 24-year-old male victim was feeding cattle using a tractor with a loader. While he was opening a bale, it fell onto him from the loader. His escape was probably blocked by hungry cattle. He died of chest asphyxia. The heavier-than-usual bale apparently caused the hydraulics to creep downward and caused the bale to topple.

—from Saskatchewan Farm Fatality Analysis, 1997-1999

Sometime it's necessary to have the loader arms elevated to work on the tractor. If the loader is not blocked (as shown in this simulated situation), unexpected downward movement of the arms could trap and injure the operator.

CASE STUDY: RIDING IN LOADER BUCKET

On April 4, 1996, a 40-year-old male died of injuries sustained when he fell from the bucket of a front-end loader and was run over by the vehicle. The victim and a coworker were riding in the bucket of the tractor, which was driving along a paved two-lane road to a parking area when the incident occurred. The victim was thrown from the bucket when the vehicle started bouncing. The operator stopped the vehicle when he saw the victim had been thrown out. A passing motorist stopped and assisted with CPR while the operator went for emergency assistance. The local police arrived shortly, followed immediately by the fire department emergency medical services. The victim was transported by ambulance to a nearby city hospital emergency room, where he was officially pronounced dead less than 45 minutes after the incident.
—from Massachusetts FACE Investigation #: 96-MA-016-01

CHAPTER 4
GETTING HITCHED

▶ *The area between a tractor and the implement can be very dangerous for the person doing the connection.*

▶ *Typical injuries in pin connections involve relatively minor scrapes, bruises, or muscle strain, but power takeoff (PTO) shaft accidents can cause severe and fatal wrapping injuries.*

Your tractor can be connected to a variety of implements for farm work, and knowing how to accomplish the connection without damage to yourself or the implement will prevent lots of pain and frustration.

Women are at a higher risk than men of being run over by tractors and other farm machinery, according to data from the U.S. National Safety Council. The increased risk is because women often assist their spouses by hitching equipment to tractors, which is an activity that exposes them to injury or death.

In a single-point drawbar connection, the tractor is backed up to the implement and a pin connects the tongue of the implement to the drawbar of the tractor, much like the connection between a car and a trailer.

Be especially careful with your fingers while hitching an implement to a tractor. They can be pinched or crushed between the tractor drawbar and implement hitch.

Use a properly sized and hardened drawbar pin when pulling heavy loads. A pin that's too small allows the load to jerk back and forth, possibly damaging the implement or tractor.

Keep implement tongue jacks in good repair. Lifting a heavy hitch by hand creates risk of back injury.

If the hitch has fallen to the ground, use jacks or levers to raise it. Trying to "muscle it up" can result in crippling back strain.

DRAWBAR CONNECTION

- When backing up the tractor to line up with the implement, do so without anyone standing in between the tractor and implement. If your foot slips off the clutch or brake, the tractor could crush the helper.
- Once the tractor is as close as you can manage to the right hitching point, set the brakes and stop the engine so the tractor does not roll away or toward you when you step behind it to insert the connecting pin.
- A 2x4 or other stout lever helps shift the implement into position without muscle strain.
- Once the hitch pin is inserted, secure it with a locking pin or clip so the connection is not unexpectedly lost during work.
- Ensure that the implement hitch jack is fully raised to its working position so that a safety hazard is not created by the jack being dragged or torn off.

- When unhitching any implement, prevent sudden movement by unhitching on level ground or setting sturdy blocks at the implement wheels. Set a large block under the drawbar jack to keep it from sinking into the ground. If the machine sinks, it becomes awkward to reattach and increases the risk of muscle strains.

THREE-POINT HITCH CONNECTION

- When backing up the tractor to line up with the implement, anyone giving directions should stand behind the implement, not between the tractor and implement. If the tractor operator's foot slips off the clutch or brake, the tractor could crush a helper standing between the tractor and implement.
- Once the tractor is as close as you can manage to the right hitching point, set the brakes and stop the engine so the tractor does not roll away or toward you when you step behind it to insert the connecting pin.

With implements mounted on the three-point hitch, backing in to make the connection is slightly more difficult because you have to line up the two connection points on the lower lift arms.

- A 2x4 or other stout lever helps shift the implement into position without muscle strain.
- Be especially careful not to get your fingers in a position where they could be pinched between the hitch arms and the implement.
- Once the link pins are inserted, secure them with locking clips so the connection is not unexpectedly lost during work.
- Before moving, test the connection by raising and lowering the implement. Ensure that any implement supports are fully raised to their working positions so a safety hazard is not created by the implement striking the supports.
- When unhitching any implement, prevent sudden movement by unhitching on level ground or setting chocks behind and in front of the implement wheels. Set a large block under the machine supports to keep them from sinking into the ground. If the machine sinks, it becomes awkward to reattach and increases the risk of muscle strains.

continued on page 62

Above: *PTO shafts that transfer tractor power to implements add another area of injury. Risk occurs both from simple pinch injuries when the yoke is being connected, and more serious wrapping injury if the shaft rotates.*

Opposite, top: *PTO shaft guards are designed to spin loosely on the driveshaft so that if a person or object accidentally touches a moving shaft, the guard will stop. This helps prevent entanglement around the shaft or in the flexible joint. The yellow shaft in the foreground has the complete guard in place, while the red shaft to the rear presents a safety hazard because part of the guard is missing.*

Opposite, bottom: *PTO safety shields are sometimes removed for convenience in hitching or break off when the shaft is in use, such as when the shaft is angled beyond normal limits. Operating a shaft like this exposes anyone to unnecessary wrapping hazards.*

Above: *Secure the PTO shield safety chains to solid parts of the tractor. To reduce the risk of wrapping injury, the safety chains keep the shield stationary while the shaft turns inside the shield.*

Opposite, top: *Hydraulic hose connections can involve a risk of eye and skin irritation or tissue damage from oil spray if a connection is attempted when there is pressure in the lines. Attempting to force the connection may result in hydraulic oil spraying out and onto the operator.*

Opposite, bottom: *When trying to track down a pinhole leak in a hydraulic hose, do not use your hand to feel for the leak, because a jet of oil could pierce your skin. Instead, pass a piece of white paper near the hose or spray the hose with antiperspirant. The leaking oil will be easily spotted on the white surface.*

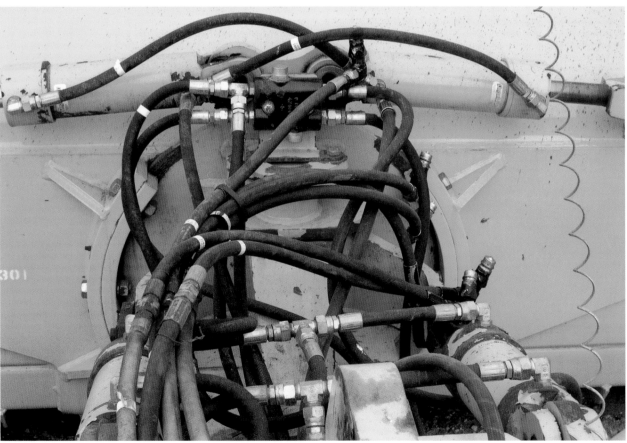

Continued from page 58

PTO CONNECTION

- Make sure the PTO selector is set to neutral. Stopping the tractor engine completely adds another margin of safety. Shut off the PTO whenever moving between areas of work.

- To prevent entanglements, footwear should be fully laced, loose clothing tucked in, and hair tied up or securely tucked under headgear.

- Do not allow anyone (including children or dogs) to remain in the tractor cab or on the operator's station while connecting the PTO. Accidental engagement of the PTO while a person is near the shaft or implement will be disastrous.

- Smear a light coating of grease on the tractor's PTO stub shaft. This allows the PTO yoke to slide on easily and prevents bruised fingers and skinned knuckles.

- With everyone clear of the connected shaft, return to the operator's station and engage the PTO at low engine revolutions to test for proper operation.

- When the PTO is disconnected, secure it to the implement so that it does not fall on the ground.

HYDRAULIC HOSE CONNECTION

- Hydraulic fluid under pressure that gets injected under the skin can cause severe, permanent tissue damage even though initial symptoms may be minor. Wear suitable eye and skin protection.

- Wipe the hydraulic connector clean before connecting to the tractor. This makes connection easier and prevents contamination from entering the tractor's hydraulic system.

- If a connector will not click into the tractor's hydraulic receptacle, pressure in the line may be interfering with connection. Relieve line pressure by loosening a joint away from the connector. Unthread the connection slowly and wrap rags around the connection to prevent oil spray and soak up any oil that oozes out. Stop the tractor engine and move hydraulic controls back and forth to relieve all hydraulic pressure within the tractor, and then reattempt the connection.

- To prevent safety problems when subsequently disconnecting the equipment, be sure all hydraulic pressure is relieved first. With hoses still connected, lower the implement to the ground or onto supports so no pressure remains in the lines. Shut off the tractor engine and move hydraulic controls back and forth to relieve all pressure. Only then should you disconnect the hoses.

RICHARD POLKINGHORNE

It was harvest time in 1983. Richard Polkinghorne climbed the ladder of the grain bin he was loading and discovered it was full. By the time he made it back to ground level, grain was spilling over the sides. He reached across the PTO shaft with his left arm to loosen the auger's belts in order to shut it down. His arm became entangled in the auger and was later amputated below the elbow.

"It was the middle of October and I had coveralls on; it was cold. It was about six-thirty or seven in the evening. I am assuming that my sleeve must have gotten caught in the power takeoff. All I was left with was my boots. The PTO took my coveralls, my coats, my clothes, and everything. So I stumbled around and just vaguely remember making it to the two-way radio in the truck. My wife, Lois, was at the base station and she heard everything firsthand and got help. Thank the Lord for small miracles!"

CHAPTER 5
MACHINERY AND IMPLEMENT SAFETY

▶ *One out of five farm accidents involve machinery and implements.*

▶ *Most of these accidents occur when the tractor is parked:*
27.5 percent occur when the machine is stopped and not running;
20 percent occur when the machine is stopped but running.

The problems of machinery and implement safety can be worse on small farms than large commercial farms because of the condition of machinery and implements used. The older equipment used on small farms is often missing the operator's manual and may also be missing many of the shields, guards, and warning labels it had when manufactured. Equipment purchased at an auction or farm disposal sale is unlikely to come with instructions or training for its proper use. It may also have suffered abuse or lack of maintenance, which makes it more dangerous in continued use.

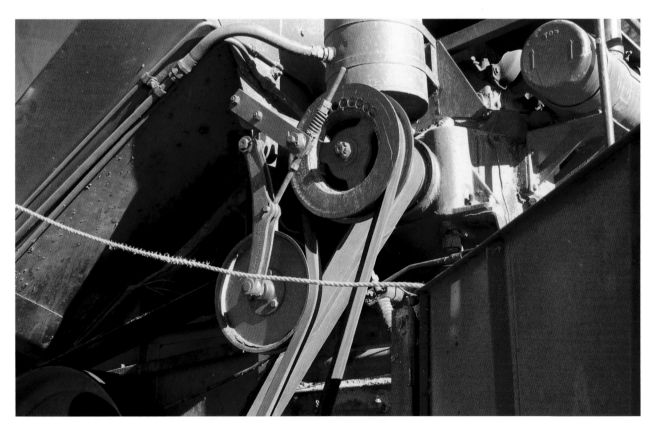

When you're around farm machinery, be aware that this means there are many hazardous mechanisms that can entangle, cut, or crush you. Take all necessary precautions to protect yourself and others.

Open drivelines are one of the main safety hazards on all machines. On the manure spreader above the chain guard is missing, while the guard is shown in place on a similar machine below. Even if the original guard is missing, the mounting points will still be there, and a shop-built guard can be made with simple sheetmetal bending techniques.

Access ladders and platforms improve work efficiency, as well as help prevent slips and falls.

When using combinations or trains of implements, such as this disc-harrow-packer combination, allow for the extra length needed when turning. Starting the turn too late and turning tightly in an effort to compensate may cause the implement to tangle with the tractor's rear wheel.

FIRST STEP TO IMPROVED SAFETY

The first step to improved safety around machinery and implements is to acquire a relevant operator's manual and refer to it for safe operating procedures. New farm equipment dealers can often access manuals for their older products. Even though the manual may not be in stock, reprints of older manuals can often be ordered. Internet auction sites like eBay or vintage tractor book suppliers such as manuals.us or ytmag.com are other places to investigate. For a search term, start with "farm equipment manual" or narrow it down with the exact name of the implement you have, such as "planter manual." In addition to private sellers, many bookstores stock farm equipment manuals, even for machines that are quite old.

Tractors and implements are heavier than cars and trucks. Before crossing any bridges, find out the permissible weight so you don't end up in the water.

Even if you don't turn up the manual for the exact make and model of your machine, a similar one can tell you a lot regarding the safety issues concerning the machine. For example, safety precautions for an old John Deere rectangular baler are similar to those of a newer John Deere rectangular baler, or even a baler of another make, even though there are differences in how the machines operate. Studying a reasonably relevant manual can improve your safety-related knowledge until you find an exact manual.

Once you have a manual, use the information to get the machine or implement as well adjusted and lubricated as you can make it before starting work. A well-maintained machine or implement is less likely to break down or plug up, and fixing or unplugging machinery is often when accidents happen.

THE ENGINE POWER DILEMMA

Ideally, when adjusting or repairing a machine, all power should be completely shut off by stopping the tractor engine or on-board engine. There is much less risk of accidental movement of the machine and subsequent injury of the operator.

Some mechanisms, such as the gathering forks in a conventional square baler, may remain under tension if the machine is accidentally plugged up during use. When the operator clears the blockage, the release of tension may cause the mechanism to move unexpectedly and injure the operator. Manuals explain these kinds of typical hazards, so make use of available knowledge to eliminate hazards.

Hydraulic power alone may not consistently hold up heavy weights for long periods of time.

Before getting underneath to work on anything held up by hydraulic rams, such as this truck box, set additional supports that will prevent the load from crushing you.

Secure any dangling hoses or wires that could trip you when getting on or off the tractor.

However, in practice, the operator often disengages the PTO or machine drive and leaves the engine idling. This is not just lack of safety awareness. For many heavy-duty diesel engines equipped with a turbocharger, shutting off the engine requires letting the engine run at high idle for several minutes to cool the turbocharger and prevent expensive damage. For this reason, many operators are understandably reluctant to shut down the engine while making what they think will be a short adjustment of the implement.

If it's impractical to completely shut down the engine, be aware of the possible risk of accidental engagement and do what you can to protect yourself. Consider ways you could reduce exposure to potential danger points, such as clearing baler blockages with a pitchfork instead of your hands or inserting a hefty wooden block in the drive mechanism to prevent accidental movement while you are near the machine. Just be sure to remove the safety block before restarting.

THE FRUSTRATION FACTOR

Many farm machinery injuries are related to blockages and breakdowns that occur during use, such as clearing hay from an improperly functioning baler. When these blockages and breakdowns occur, the operator is likely to be in a frustrated or agitated frame of mind, especially if the problem is occurring repeatedly or has not been resolved by normal measures.

When frustration builds up, reasoned judgment often goes out the window. Alternatives that are much less risky but might take a little longer are overlooked or quickly rejected, even though the delay they bring is much less than the delay entailed by an injury. A person who thinks he or she is behind on the job may give little thought to how much further he or she would be missing out as the result of an injury.

Understanding that stress and frustration disrupt your usual thought processes is a reminder to prepare countermeasures in advance. Consider it a situation similar to a pilot's checklist when he or she encounters a stress situation. Farm-stress checklists can be as simple as taking a deep breath for 30 seconds or so to consider the options before taking action or to always shut off the machine before making repairs or adjustments.

When you're frustrated, a "just get 'er done" focus can more easily lead to a work-stopping or life-ending accident. In times of frustration, a personal checklist of countermeasures can be an alternative that improves both short- and long-term personal safety.

Years of jumping off tractors can lead to chronically sore "farmer's knees." Dismounts facing the tractor (top) put less shock on your knees than hopping off face-out (above).

GENERAL SAFETY PROCEDURES

1. Wear suitable protective equipment such as heavy gloves, eye protection, hearing protection, and boots with nonslip soles.
2. Observe any warning labels and apply their advice to your work.
3. Before attempting to unplug, clean out, adjust, repair, or lubricate the unit, turn off the source of power and physically block any parts that may suddenly move when an obstruction is removed.
4. Once power is off, wait for the mechanism to come to a complete stop before approaching the machine.
5. Do not allow children or pets to remain in the tractor cab or operator's station. They could accidentally engage the machine drive or PTO.
6. Secure any dangling clothes, hair, or jewelry that could get tangled in the machine.
7. Never attempt to engage or disengage a v-belt drive by pulling on the belt with your hands.
8. If the machine is held up by hydraulic power, set blocks to prevent unexpected movement before you crawl underneath the machine.
9. Parts that move during operation may be very hot to the touch—test them lightly at first to avoid burns.
10. When you are bent over or lying under the machine, get up slowly to avoid banging your head or back.
11. Before you move a machine or engage the drive, be sure you know where any bystanders, especially children, are located.
12. Avoid moving in reverse unless you can be certain there is no one behind you.

The rapidly turning screw (flighting) can quickly sever body parts. Belt and PTO drives pose additional risks.

MACHINE AND IMPLEMENT SAFETY HIGHLIGHTS

The following alphabetical listing of farm machines and implements reviews some of the key dangers and safe procedures to follow in use. When you purchase a new or new-to-you machine, reviewing the operator's manual is also highly recommended for both your safety and your family's safety.

CASE STUDY: AUGER HAZARDS

A 65-year-old man working for a farmer during harvest was unloading a gravity flow wagon of corn into a partially unguarded floor hopper. His right leg was caught in the auger at the bottom of the hopper and severed his leg at the mid-thigh region. The man was found lying unconscious next to the hopper and had significant bleeding. The auger was still running at the time he was found by the farmer's mother. First responders were notified and arrived within a few minutes, but rescue efforts were ineffective due to the large amount of blood loss. The man was taken to a local hospital where he was pronounced dead.

A chewed-up broom handle was found next to the victim. The 2-foot-wide floor hopper was covered with a steel grate to allow vehicles to ride over the hopper. Nine of these bars were removed during the time of the incident, leaving an 18-inch section of the auger exposed. Since the wagon was empty, it appears that he was finishing the unloading and sweeping up spilled corn on the floor when his foot slipped into the uncovered hopper and was caught by the moving auger.
—from Iowa FACE Report #95IA047

70

Auger

The rotating screw (flighting) carries feed, grain, or other material up the auger tube. It also creates a crushing or cutting hazard if hands or feet get caught between the screw and the tube. Grain auger injuries often involve relatively undertrained women and youths who are assigned the task of unloading grain trucks while older males run the harvesting machinery. Auger intakes should have grates with openings only wide enough for the material, not wide enough for hands or feet.

- If an object, such as a cap, falls into the auger, do not reach in to grab it.
- In windy conditions, be careful about raising the auger to full height. Strong winds can tip it over sideways.
- If the auger is raised and lowered with a cable winch, do not let the crank handle spin freely and do not try to stop it with your hand if it does get away.
- Contact with an overhead power line can cause an electrocution injury. Check overhead whenever lifting or transporting the auger.

Above and right: *Augers equipped with a permanently attached hopper are more convenient to use, and they allow inclusion of bars or a grate to keep body parts away from the flighting.*

CASE STUDY: AUGERS AROUND POWER LINES

A 51-year-old male dairy farm owner was electrocuted when the grain auger he was moving came into contact with a 7,200-volt power line. The victim was relocating a 41-foot grain auger from one storage bin to another when the elevated end contacted the overhead service line. The victim was pronounced dead at a local hospital.
—from Colorado FACE Report #90C0042

During transportation of the auger, contact with overhead wires is also a risk. Scout your transport area and lower the auger to transport position whenever practical.

Never start the auger engine by pulling on the drive belt. Your hand could get caught and pulled into the drive pulley.

Baler (rectangular)

- Review the operator's manual for recommendations on how to pull the baler so that material enters correctly to make a good bale (e.g., strokes per bale and string tension). If you learn the techniques to create a good bale, you'll be less exposed to hazards trying to adjust or unplug the machine.
- Never feed material into the baler by hand, such as rebaling a broken bale. Spread the material out on the ground and let the baler pickup gather it.
- Do not try to pull twine out of the knotter while the baler is running.
- After the PTO is off, wait until the flywheel has completely stopped turning before approaching the machine.
- Gathering forks, plungers, and knotters in a conventional square baler may remain under tension if the machine is accidentally plugged during use. When clearing the machine, use blocks in the mechanisms so that the release of tension does not cause unexpected movement that could entangle you in the machine.
- Use only the correct type of shear bolts in the baler. If shear bolts fail repeatedly, fix the cause and not just the symptom. Refer to the manual for the location of points protected by shear bolts.

Baler (round)

- Review the operator's manual for recommendations on how to pull the baler so that material enters correctly to make a good bale (e.g., weaving slightly from one side of the windrow to the other). If you learn the techniques to create a good bale, you'll be less exposed to hazards trying to adjust or unplug the machine.
- Never feed material into the baler by hand, such as when rebaling a

Risk of injury includes entanglement in the PTO shaft or pickup, getting cut by the knives, or crushed by the compression plunger and tying mechanisms.

broken bale. Spread out the material on the ground and let the baler pickup gather it.
- Use only the correct type of shear bolts in the baler. Using stronger bolts will cause other parts of the machine to break.
- Stop occasionally to clean chaff off the baler and check for proper belt tension on variable-chamber balers.

Chisel plow

- Engage mechanical safety locks or use strong blocks whenever it's necessary to climb under the machine.
- If the chisel plow plugs up with corn stalks or straw, raising the implement and backing up may clear the blockage without the operator climbing under the plow to clear the blockage.

Shear bolts (at locations such as the one highlighted above) are special bolts that protect the baler by failing in case of a stoppage in the machine. If stronger bolts are substituted, there will be more expensive damage inside the baler and you will be exposed to increased hazards during repairs.

If you need to clear plugged material from the pickup, stop the PTO power to the baler and block the gathering forks so they cannot accidentally move when the blockage is removed. Remove the block before restarting.

Risk of injury comes from entanglement in the pickup and abrasion or crushing in the chamber. Additional hazards are entanglement in the PTO shaft and fire due to belt and bearing friction.

Carry a large fire extinguisher on the baler or tractor. Small pieces of dry material may easily catch fire if belts are slipping or bearings are overheating.

If the bale ejection door at the rear is open, engage the safety lock or close the hydraulic valve before moving underneath the rear door.

TONY POTOREYKO

"I'm not the type of guy that takes chances, like taking shields off and things like that. That's one of the things that really frustrated me. You practice safety to the best of your ability and you're the one who gets caught. After the accident, somebody said there was a root sitting behind the back of the baler and they thought I tripped and reached out to catch something. A universal or automatic shutoff for the tractor engine would have helped. The accident would have happened, but the damage would have been very small because within seconds I could have shut the tractor off. When you're alone in the situation and there's no way to shut the equipment down, it's a real hellhole."

On August 6, 1993, Tony Potoreyko was entangled in the rear end of a round baler. He can't recall anything between dismounting the tractor to check on the baler and regaining consciousness in the baler. His right arm was stuck between the roller and a rotating belt. Suddenly he remembered he had a utility knife in his jeans pocket. Knowing the knife was his only hope of getting out, he used his left hand to cut at the 8-inch belting as it stripped the flesh from his right arm. Eventually, he managed to cut through the belt and free himself. Reconstructive surgery saved his arm. Tony estimates he has about 40 percent of its use now.

DANGER

DON'T TAKE CHANCES!

To avoid injury or death by being pulled into the machine: Do not attempt to feed crop or twine into baler or unplug feed area WHILE BALER IS RUNNING. The baler feeds material faster than you can release it. Disengage PTO and shut off engine.

Below: *Wear heavy gloves when changing shank tips. Lacerations may occur when pushing on the heads of the plow bolts that hold the tips to the shanks. If using an air wrench, the head can spin suddenly and severely cut the end of a finger.*

- Try operating at lower speeds so that the crop residue has an easier time flowing through the machine. If the problem persists, the field may be too damp to work with this implement. Working the field with tine harrows at 6 to 8 miles per hour can help break down residue into smaller pieces.
- Stand up slowly to avoid banging your head or back.
- If you are welding cracks on frame members of the chisel plow that are made of square or round tubing, hot gases can build up inside the tube and cause an explosion. Before starting the weld, make sure there is a vent for gases to escape freely.
- When transporting the cultivator with its wings folded up, watch out for overhead power lines.

Combine

- Replace operating shields and guards after maintenance or adjustment.
- Adjust or lubricate the combine only when it is completely stopped.
- Adjust sickle sections only when the reel and header drive is off.
- Block the header or reel securely before climbing underneath.

- Keep ladders and platforms clear of grease and obstructions.
- Keep cab windows and mirrors clean.
- When backing up, take a moment to make sure no one is behind the combine. Due to the size of the machine, it is very hard to see to the rear.

continued on page 80

The main safety concern with this implement occurs when the operator climbs underneath to make adjustments or repairs while the implement is stopped. If the machine is held up only by hydraulic force and isn't mechanically blocked, it can accidentally lower and trap the operator underneath.

Hazards associated with the combine include contact with various cutters, augers, conveyors, beaters, impellers, drive belts, and roller chains, plus the dangers of falling from the tall machine. It is important to follow maintenance and lubrication schedules because a poorly maintained machine is more likely to perform badly and bring the operator into contact with hazardous areas during adjustment or repair.

BRUCE KISTNER

On October 7, 1980, Bruce Kistner started baling early in the day. By 10:00 a.m. it was too dry to make a good bale, but he kept going. Although it was his 19th wedding anniversary, he couldn't take the day off. Cattle were getting into the crop and he wanted to finish the baling. When the machine's pick-up plugged, he grabbed the straw to unplug the machine. His hand was pulled in between two rollers and held tight until it burned off.

"It was bad fall weather for harvesting. I had combined for about two nights and two days by myself. I was tired. You know, I could watch the combine plug and I wouldn't even stop until it was plugged. I finished at midnight and went out after that to bale. Then I started baling again about five o'clock in the morning. I wasn't looking for anything so I guess I was just overtired, that's all. If I could go back and do it over again, I suppose I would shut the baler off, open the gate on the back, and dig the straw out from behind by hand. You could pull it out easy by hand with the baler open."

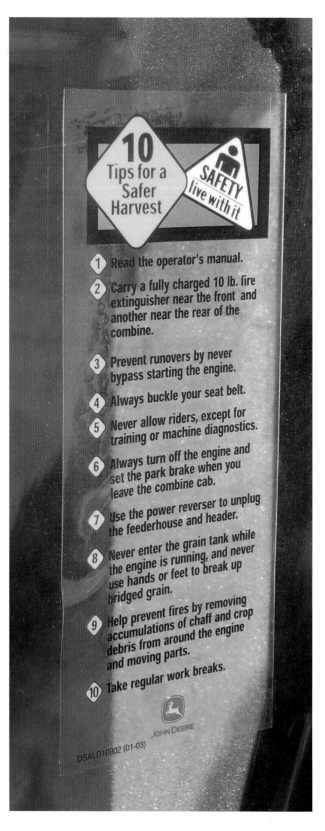

This sticker helps keep safe operating methods front and center while you are on the combine. This sticker can be ordered at no charge from John Deere dealers and is part number DSALD12406B.

Older, gasoline-powered combines, such as this model, are a greater fire hazard than diesel combines because of the higher operating manifold heat and exhaust sparks associated with gasoline engines. Regularly clean dust and chaff from the engine bay and radiator area.

Every combine uses a large number of belts and chains to transfer power from the engine to the transmission and threshing mechanisms. Keep the drive shields installed and close them after completing service or repairs.

In difficult harvest conditions (e.g., tough straw, heavy weeds, or uneven crop) the infeed area of the combine (knives, pickup, table auger, and/or feeder chain) can easily become blocked and require the operator to pull out material by hand. Be certain that both header and threshing drives are fully disengaged before entering the table area.

Continued from page 76

- Make sieve adjustments only when the machine is completely stopped. Contact with the nearby straw chopper or spreader can cause severe injury.
- Stay out of the grain tank unless the combine is shut off and no one is at the operator's station.
- Keep clear of the rear of the machine when the combine is running because straw is propelled out of the machine at high speed.
- During transport, empty the grain tank to reduce weight and lower the combine's center of gravity. Move the unloading auger to the transport position to avoid striking power poles and other objects.
- With any combine, be particularly careful about fire safety when combining sunflowers. A fine-grained, highly flammable dust from the sunflower heads accumulates on the combine. Clean it off regularly.

Conveyor (hay bale)

- Secure any dangling clothes, hair, or jewelry that could get tangled in the machine.
- Never attempt to start the engine by pulling on the drive belt with your hands.

continued on page 84

Do not attempt to install a v-belt by hand-directing it onto a rotating shaft. The belt can pull your hand into the pulley faster than you can react. Loosen the belt tensioner to allow the belt to slip on and then retighten the tensioner.

Falls can be more hazardous than you think when working on high areas of the combine without access platforms and guard rails. Job safety statistics show that in a fall of 11 feet or more, 50 percent of victims will die.

CASE STUDY: COMBINE INJURIES

1) An 84-year-old farmer was killed while cleaning dirt out of an old combine header. The victim was working alone and combining beans in a remote field. He had purchased a used, early 1970s self-propelled combine the previous year. This was the first time he used the combine in the field. He had difficulty operating the combine because it did not separate the beans properly and the cutting height was too high or too low at times.

The left side of the combine header had become clogged with field dirt because it traveled too close to the ground. The farmer had apparently disengaged the header using a control lever in the cab and climbed down to clean dirt off the sickle and auger. He left the engine running because it had a bad battery.

Apparently the header slipped into gear on its own while he was cleaning the dirt. The reel and auger had thrown him around to the right side and pulled his left leg into the auger. The chain driving the reel and auger broke while he was caught between the reel and auger with the sickle still running. He was found unconscious, lying on the sickle with a significant leg injury and loss of blood. An emergency crew used the Jaws of Life to cut through the reel to remove the victim, who died on the way to the hospital.
—from Iowa FACE Report #95IA035

2) A 60-year-old farmer died from injuries sustained when he was run over while servicing a combine. Scene conditions indicated that the combine was left running while the victim attempted to add hydraulic fluid. During this maintenance, the combine suddenly moved forward. The victim was run over by the rear tire. After circling the field, the combine ran over the victim a second time. There were no witnesses to the incident.
—from Oklahoma FACE Report #03OK03201

3) A 74-year-old Iowa farmer fell to his death while getting his combine ready for an auction sale. He needed to add antifreeze to the radiator located on the right side of the engine on top of the combine. To perform this task he needed to climb on top of the combine using the stationary service ladder on the right side of the machine and stand on the top steps of the ladder or on the maintenance platform behind the engine area. There are no guardrails at this location on this machine. The events were not witnessed, but it appears the man fell from the ladder or platform and received fatal head injuries while falling to the concrete patch in the narrow space between the combine and the crib wall.
—from Iowa FACE Report #98IA048

Since combining always involves highly flammable dry straw and chaff, keep a large fire extinguisher handy. During daily greasing and inspection, brush away buildups of dry material near any parts of the combine that can become hot, such as exhaust pipes and bearings.

Above: *The open conveyor on this machine presents an entanglement hazard, because in order to place a square bale on the conveyor, operators typically stand quite close to the chain. During transportation of the conveyor, contact with overhead wires is also a risk.*

Right: *Electric motor drives are often used on bale elevators because the machine is used near barns with electric plug-ins, and the drive is quiet and easy to start. Ensure that connections to the motor are kept in good repair to prevent dangerous transfer of current to the highly conductive steel frame of the elevator.*

Opposite: *As convenient or devil-may-care as it might look to do it, never attempt to ride up the conveyor on a bale. A small slip leaves feet and hands dangerously close to the chains.*

Continued from page 80

Conveyor (grain)

- Conveyor intakes should have grates with openings only wide enough for the material, but not wide enough for hands or feet.
- If an object, such as a cap, falls into the conveyor, do not reach in to grab it. Stop the conveyor first and make sure it is completely stopped before retrieving the object.
- In very windy conditions, be careful about raising the conveyor to its full height. Strong winds can tip it over sideways.

The fast-moving belt that moves grain up the conveyor tube also presents an entanglement or abrasion hazard if hands or feet get caught between the belt and the tube.

If the conveyor is raised and lowered with a cable winch, do not let the crank handle spin freely and do not try to stop it with your hand if it does get away.

Corn picker

Pickers are frequently plugged if ground speed is too fast or slow, which presents a major entanglement risk if the operator attempts to clear out plugged material without stopping the picker and shutting off the tractor or picker drive.

CASE STUDY: CORN PICKER HAZARDS

On November 22, 1995, a 52-year-old farmer died when he was caught in the corn picker he was operating. He was a full-time farmer and had been involved in the family farming business all of his life. When he did not return home at the expected time, his wife and son went looking for him. They found him entangled in the rotating shaft of the corn-picking machine.

Although there were no witnesses to the incident, the family speculates that the victim went to see if the husking bed was empty and left the equipment running, as he usually did. As he stood up on the equipment and reached across the husking bed near the rotating shaft of the ear forwarder that moves the corn through the husking bed, his unzipped jacket likely became caught in the moving parts. As the shaft continued to rotate it pulled his arms and upper torso under and around the shaft.

The operator's manual for this equipment and safety stickers present on the equipment state that adjustments to the machine should never be attempted while the machine is in operation.

—from Kentucky FACE Report #95KY126

Above: *Snapping rollers travel at about 12 feet per second, which means a farmer holding onto a stalk 2 or 3 feet away from the mechanism has less than a half-second to react.*

Opposite, top: *Stop power to the mechanism before reaching anywhere into the machine to clear blockages.*

Opposite, bottom: *Don't forget that the unloader is higher than other parts of the machine, whether it extends to the rear on a grain-corn machine or to the side on the sweet-corn picker. The extra height means potential risk of contact with overhead electric wires.*

Cultivator

- Engage mechanical safety locks or blocks whenever it is necessary to climb under the machine.
- Wear gloves when changing shanks. Lacerations may occur when pushing on the heads of the plow bolts that hold the tips to the shanks. If you are using an air wrench, the head can spin suddenly and slice your finger.
- When you are getting up, stand up slowly to avoid banging your head or back.
- If you are welding cracks on frame members of the chisel plow that are made of square or round tubing, be aware that hot gases can build up inside the tube and cause an explosion. Before starting the weld, make sure there is a vent for gases to escape freely.
- When transporting the cultivator with its wings folded up, watch out for overhead power lines.

CASE STUDY: SHUTTING OFF POWER TO IMPLEMENTS DURING SERVICE

A 42-year-old farmer was fatally injured when he attempted to adjust the spacing on the cultivator he was using. He had stopped the forward motion of the tractor but left the tractor running. He raised the cultivator in order to adjust the hilling wings. The cultivator was equipped with a hydraulic-powered weeder bar that was powered by the tractor's PTO. As he exited from under the cultivator, his jacket caught on the connecting bolt of the rotating weeder bar and twisted to the point of strangulation. When coworkers found the victim they notified the county sheriff and coroner. He was pronounced dead at the scene.

—from Colorado FACE Report #92C0025

Above and opposite: *The main safety concern with this implement comes when the operator climbs underneath to make adjustments or repairs while the implement is stopped. If the machine is held up only by hydraulic force and isn't mechanically blocked, it can accidentally lower and trap the operator underneath.*

Disc

- Disc blades have sharp edges, so be very careful when working or making adjustments around them.
- Engage mechanical safety locks or blocks whenever it is necessary to crawl under the machine.
- When you are getting up, stand up slowly to avoid banging your head or back.
- If you are welding cracks on frame members that are made of square or round tubing, be aware that hot gases can build up inside the tube and cause an explosion. Before starting the weld, make sure there is a vent for gases to escape freely.
- When transporting the disc with its wings folded up, watch out for overhead power lines.
- If the disc is mounted on a three-point hitch, turning the tractor swings the disc the other way, so turn carefully to avoid hitting objects.

When removing a gang of discs to change pans or bearings, be aware that the gang can be very heavy. Lower the gang using jacks or ropes attached to the frame.

Drill (seed)

- Lower the grain drill to the ground before servicing or adjusting the drill, such as when removing bags from the openers during calibration or changing soil openers.
- If seed or fertilizer boxes are filled, the drill must be hitched to the tractor or otherwise supported to prevent tipping.
- No riders are allowed while the grain drill is being operated. The catwalks at the rear of the drill are for use only while the drill is stopped. A sudden jolt or bump while moving could cause the rider to fall off.
- When filling the drills, be sure the catwalks are clear of dust, grease, and obstructions to reduce the risk of slips or falls.
- Be careful during turns because if the drill catches on a tractor tire it could flip the drill over sideways.
- When transporting the drill, be sure any transporting device (end wheels) are properly positioned and secured with locking pins.

CASE STUDY: SEED DRILL STABILITY

A 71-year-old farmer was killed in an accident on the family farm in 1960. The accident happened when he was oiling the wheels on his grain drill, which wasn't hooked to the tractor. The drill turned over on him and killed him instantly when the grain shifted.
—From unit history of USS Princeton (http://freepages.military.rootsweb.com/~cacunithistories/USS_Princeton.html)

Placement of the seed box more or less directly above the pivot point (wheels) means the weight of grain can make the end-wheel type drill subject to forward or backward tipping if the drill is unhitched with full seed boxes.

Packer-type drills are less subject to tipping because the weight is borne by dolly wheels at the front and the packers at the rear. Periodically inspect the catwalks located at the rear for operator access. When you fill the seed boxes, your attention will be focused on the job, which makes it easy to slip from a catwalk in poor repair.

CASE STUDY: MIXER MILL WITH UNGUARDED PTO SHAFT

A 64-year-old Iowa cattle farmer was killed when his hooded sweatshirt became entangled in the power shaft of a roller mill that was grinding corn. The roller mill was powered by a 15-kilowatt (20-horsepower) electric motor via two V-belt transmission pulleys and a PTO shaft approximately 50 mm (2 inches) in diameter. This shaft was connected to the mill-power intake shaft by a U-joint on the side of the mill. A protective master-guard over the U-joint had evidently been removed and left this joint exposed. There were no witnesses to the event, but from the circumstances it appeared that the man possibly slipped and/or fell onto the rotating shaft. There was no obvious reason for him to be in that position near the roller mill. There was an oil can on the floor, which could have been used at the time for oiling chains, but the chains were on the other side of the mill. The controls for the incoming and outgoing corn were on the other side of the mill as well. There were nine sacks of feed next to the door, close to the mill, and there were several objects on the floor and around the space next to the mill where the victim would have been standing. It is possible that he stepped on or was caught by something, which made him lose his balance. The farmer died from cervical fractures.

—from Iowa FACE Report #01IA005

Feed mixer-mill

- Make sure the PTO shaft has all guards and shields installed, and keep clear of the shaft in operation.
- Avoid overloading the grinder, which can result in plugging that has to be cleared by an operator.
- Change mill screens or clear plugs only after the hammers have come to a complete stop.
- Shoveling grain directly into the hammer mill can cause the shovel to come in contact with the hammers.
- Slippery, uneven ground or horseplay around the intake hopper may cause the operator to make hazardous contact with the machine.
- When you are guiding the spouts to a bin or truck, take care to prevent pinched fingers.

Forage harvester

- Block the header securely before climbing underneath.
- Know and follow the manufacturer's procedures for sharpening the knives. Stand in the correct position and do not lean over the sharpening mechanism.
- When sharpening, wear the recommended personal protective clothing, including eye protection.
- Before opening any guards or shields, allow the cutting mechanism and drives to come to a complete stop.
- Know and follow the correct procedures for clearing a chute blockage.
- Use the reversing drive to clear blockages. Before you attempt to clear any blockage by hand, make sure the cutter-head is stopped, the PTO is disengaged, and no one is near the tractor controls.
- Follow maintenance and lubrication schedules. A poorly maintained machine is more likely to perform badly and cause the operator to make adjustments.

Safety hazards in a portable grinder-mixer mill involve crushing by the hammer mill section, cuts by the mixing auger, or entanglement in the slinger or PTO drive. Breathing dust when cleaning the bin is also a hazard.

Forage harvesters used on small farms are often older pull-type models rather than the large self-propelled units that are similar to combines. Typical hazards on the pull-type machine include entanglement with the PTO or moving drive mechanisms, contact with the exposed rotating cutter-head while sharpening, getting trapped or injured by a falling header, or falling while adjusting the discharge spout.

Forage wagon

- Before approaching the machine for inspection, adjustment, or repair, disengage the PTO and allow the unloading beaters to stop completely.
- If it is necessary to climb inside the machine, be cautious about slippery areas that could cause falls that lead to entanglement in the conveyor (raddle chain) and/or unloading beaters.
- When disconnecting the forage wagon, block the wheels. Use a drawbar jack set on a block to prevent the jack from sinking into the ground.

JOE STACHURA

On January 15, 1985, Joe Stachura was trying to unplug a feed mill driven by a PTO shaft from the tractor. The mill was clogged with a frozen bale. Joe had unplugged the mill while it was running many times before, but this time his left arm became entangled. Due to the extent of tendon, nerve, and blood-vessel damage, his arm had to be amputated.

"What happened was that the grate between my hand and the hammers slipped. The hammers took the big mitt I had on and rolled it under, catching the index finger of my left hand. It was all over in the twinkling of an eye or a blink. Maybe I shouldn't have used the frozen bales. I could have thrown them aside and just used dry straw, which goes through, no problem. When something isn't going right and stress is building, take a break—two, five, or ten minutes. Have a drink, lunch, or just look at something else for a while. Better to take a minute than to lose an arm or leg."

Hazards include entanglement with the PTO shaft or unloading beaters and getting struck by a rolling or tipping wagon. AGCO Corporation

CASE STUDIES: FORAGE WAGON HAZARDS

1) A 17-year-old farm youth died after he became entangled in the unloading beaters of a forage wagon. The victim was working near a barn and used a tractor and a PTO-driven forage wagon to deliver forage to cows in a barn. One of the conveyor chains on the floor of the wagon was broken, and as a result, the forage would not move forward toward the unloading beaters. When the youth leaned over the top beater to determine if he needed to shovel any more of the forage from the wagon, his jacket got caught in one of the tines of the top beater.
—from Minnesota FACE Report #05MN010

2) A 22-year-old farmer was killed while cutting corn silage on his farm. A filled, disconnected wagon was on a slight slope and it slowly began to creep forward. The victim first jumped on the tongue, expecting it to dig into the ground and stop the wagon, but it continued to move. He then picked up the tongue and tried to turn the wagon uphill to stop it from rolling. At this point, he was walking/running backward in tall grass. His brother yelled to him to get out of the way, but he apparently slipped and his pant leg was caught by the wagon's running gear. The silage wagon continued to roll several hundred feet downhill, dragging the victim.
—from Iowa FACE Report #02IA050

This page and next: *Since there are no moving parts on this implement it is generally safe. The main concern is that harrows are awkward to handle and are sometimes dropped, which can scrape or crush legs and feet.*

CASE STUDY: MANURE SPREADER ENTANGLEMENT

A 24-year-old male farmer died after becoming entangled in the unguarded rotating driveline shaft of a manure spreader. The spreader was connected to a tractor that powered the spreader with a PTO shaft. The victim was working alone in the barnyard, replacing a bolt on the shaft. He apparently had completed this task and was standing on ice-covered soil near the rotating driveline. Then he either slipped and fell onto the driveline or his clothing was caught and pulled in by protruding parts of the rotating shaft. He was spun around the driveshaft and portions of his clothing were entangled on the driveshaft and torn from his body. When the young farmer did not return to the farmhouse, his wife approached the site of the incident and found him entangled on the driveline. The tractor engine was not running. She called to the victim's brother, who was working in the barn, and he freed the victim by cutting the tightly tangled clothes. The brother summoned EMS, while the victim's wife began CPR. EMS responded within several minutes. The coroner's office was contacted and pronounced the victim dead at the scene.

—from WI FACE Report 99WI012

Harrow

- When a number of harrow sections are attached to a wide hitch bar, make wide, sweeping turns to prevent the ends of the hitch from striking and riding up the tractor wheels.
- Handle harrow sections with care to prevent them from falling on your feet and legs.
- Wear gloves when cleaning straw or stalks out of plugged up harrows. The densely packed material may have picked up wire or other bits of metal that can injure your hands.

Manure spreader

- PTO side shafts and the side shaft that transmits power to the beaters on conventional spreaders are key hazard areas. Keep them shielded and re-install shields after repairs.
- Always shut off power to the spreader before scraping the sides or entering the spreader box.
- To help prevent entanglement in shafts, chains, beaters, or augers, secure any loose clothing or hair.

Hazards include entanglement with the PTO shaft or unloading beaters. Keep all guards and shields in place to prevent entanglement.

If you need to crawl underneath the moldboard plow to make repairs or adjustments, make sure the machine is held up with sturdy blocks. If the hydraulics fail, the implement can lower and trap you underneath.

Moldboard plow

- Engage mechanical safety locks or blocks whenever it is necessary to climb under the machine.
- When you are getting up, stand up slowly to avoid banging your head or back.
- If you are welding cracks on frame members that are made of square or round tubing, be aware that hot gases can build up inside the tube and cause an explosion. Before starting the weld, make sure there is a vent for the gases to escape freely.
- If the plow is mounted on a three-point hitch, turning the tractor swings the disc the other way, so turn carefully to avoid hitting objects.

Mower

- Before working anywhere near the conditioners at the rear of a mower-conditioner, disengage the PTO and let the mechanism stop completely.
- Stones or other debris can be propelled out of the mower at high speeds, so keep bystanders away.
- If working under the mower, use blocks to prevent the mower from falling on you. Hydraulic lifts alone are not sufficient for holding up the machine while you are under it.

Hazards include lacerations from the cutter bar or blades; entanglement in the PTO, reel, or conditioning mechanism; and crushing by the weight of the mower.

CASE STUDY: MOWING ACCIDENTS

1) A 56-year-old grain farmer was killed when the seat of his tractor fell off while he was driving through a washout area in a cornfield. The man was driving a 1950s-model two-cylinder, tricycle-type tractor and pulling a rear-mounted rotary mower. His other tractor was in the shop, so he used this older tractor. He had recently rented the field and was working it for the first time, mowing corn stalks and preparing the field for spring planting. While mowing, he drove into what appeared to be a washout from a tile standpipe. The force of the jolt knocked the seat and the farmer off the tractor. The man fell into the washout area and the mower ran over him, causing fatal head injuries. He was found dead the next day.
—from Iowa FACE Report #98IA056

2) A 76-year-old farmer was killed while working on a rotary mower in his machine shed. He raised the mower to a sufficient height to work underneath it, but he did not provide support or blocking for the mower. While he was lying on the floor working on the mower, the tractor hydraulics were leaking and the mower was slowly coming down. He was probably aware of the hydraulic problem because he had been working with this same tractor and mower for several years. At one point, he apparently tried to roll out from under the mower, but became trapped between the right-rear wheel of the mower and the mower deck. The mower continued to come down and pinned him to the floor. The official cause of death was from suffocation.
—From Iowa FACE Report #98IA56

On some mowers, the hydraulic motor that needs to be connected to the tractor PTO is very heavy. To avoid back strain and pinched fingers, lift with care or have someone help you.

Shut off the power to the machine when clearing material that gets wound around these rollers.

With disc or drum mowers, always operate with the safety curtain in place to prevent debris or bits of stalk from being ejected at high speed. If debris flies out of the front, reduce PTO speed and increase tractor forward speed.

Completely disengage power and confirm that the blades have come to a full stop before working on any mower. The blades move at a high speed even if the PTO is turning at tractor-idle speed, and it can do severe damage to you.

Planter

- Engage mechanical safety locks or blocks whenever it is necessary to climb under the machine to replace discs or bearings.
- When you are getting up, stand up slowly to avoid banging your head or back.
- If you are welding cracks on frame members that are made of square or round tubing, be aware that hot gases can build up inside the tube and cause an explosion. Before starting the weld, make sure there is a vent for gases to escape freely.
- If the planter toolbar is mounted on a three-point hitch, turning the tractor swings the planter the other way, so turn carefully to avoid hitting objects.
- When transporting a large planter with its wings folded up, watch out for overhead power lines.

Mowing operations near public roads involve the risk to others as rocks or garbage are ejected at high speed and strike passing cars. For this reason, a flail-type mower may be preferred over the conventional rotary mower.

During repairs, if the machine is held up only by hydraulic force and isn't mechanically blocked, it can accidentally lower and trap the operator underneath.

CASE STUDY: PLANTER SERVICE HAZARDS

A 55-year-old farmer died while changing a tire on a planter. The man had been planting corn earlier in the day without problems. However, when he moved the planter to another field, he noticed that the second road tire on the right side of the planter was flat. He stopped in a pasture and began to change the tire. He did not brace up the planter nor use the mechanical safety catch adjacent to the hydraulic cylinder, which is designed to keep the cylinder extended during maintenance.

He crawled under the planter and loosened the bolts on the wheel, but he could not get it off. He then tried to relieve ground pressure on the flat wheel by loosening the hydraulic fitting connected to the wheel hydraulic cylinder. He was obviously not aware that the hydraulic lines to both wheel cylinders were connected. He expected the outside wheel cylinder on that side of the planter to hold up the machine while he relieved ground pressure on the inside wheel. However, when the hydraulic line fitting was off, hydraulic fluid immediately gushed out and the entire right side of the planter fell on the victim. The man was killed instantly due to a fatal head injury, but he was not discovered until the next day by an employee of his farm.

—from Iowa FACE Report # 95IA041

Posthole auger

- Posthole augers are designed for one-person operation from the tractor seat. No person should be allowed within 25 feet of the digger when the power is engaged.
- Do not operate the auger when another person is in contact with any part of the frame, such as putting his weight on the frame to improve penetration of the soil.
- Never replace the shear bolt or auger retaining bolt with one longer or stronger than those supplied or specified by the manufacturer.

Post pounder

- Never put your hand on the top of the post as a guide. When the heavy, fast-moving weight hits your hand, the damage will be quick and severe.
- Wear hearing protection. The repetitive clang of that big weight hitting the post happens near your head and can damage your hearing.
- Wear gloves and eye protection to prevent injury from splinters.
- To help avoid tip-overs, lower the pounder weight before transporting the machine, especially on rough or sloping ground.

The rotating auger can entangle a person standing nearby, especially when the auger oscillates as it is lifted out of a hole with the tractor power still engaged.

CASE STUDY: POST HOLE AUGER ENTANGLEMENT

A female teacher was flown by helicopter ambulance to Baylor Medical Center in Dallas after her left arm was caught in the auger of a tractor-mounted post-hole digger. It is believed that her glove was caught by the digger and her arm was twisted between the wrist and elbow. A man operating the tractor summoned help and an ETMC ambulance from Gilmer responded. A surgeon was able to save her arm, and reports indicate she would recover and retain most of its use.

—from Gilmer, Texas "Mirror," Internet Edition, Friday July 22, 2005

Potato harvester

• Potato harvest often occurs late in the year during muddy and/or frosty conditions, which create many slip and fall hazards.

• Ensure each worker is securely positioned before moving and that the guardrails are installed around the sorting area.

• Keep walkways clear of mud that can cause slips, falls, and strains.

Rake (hay)

• Before approaching the rake for inspection, adjustment, or repair, disengage the PTO and allow the tines to stop completely.

• If it is necessary to lean inside the rake to clear blockages, be cautious about slippery areas, which could cause falls that lead to entanglement or getting poked by the tines.

Hazards include crushed fingers, splinters from the wooden posts, and hearing damage from the sound of the hammer hitting the post.

Hazards include entanglement in the rake or PTO on powered rakes and getting poked by walking too close to the tines on wheel rakes. Shut off the PTO before working on or near powered rakes because it is possible to be struck in the head by a rotating reel or be pulled into the rake when tines catch loose clothing.

CHAPTER 6
DUST, MOLD, GASES, AND CHEMICALS

▶ *Farmers account for more than 30 percent of adults disabled by respiratory illnesses, yet a large percentage of farmers are nonsmokers.*

Living out in the country is generally associated with being in clean, fresh surroundings. For the most part, this is true, but awareness of hazardous and polluted periods and occasions will help keep you and your family alive and healthy. One study of 1,000 farmers found almost four times the likelihood of high pesticide exposure among those with fatalistic opinions. Farming is more dangerous than jobs in industry or manufacturing and accidents are just one of the occupational hazards of farming that you must accept if you are going to be in

the business. To make a profit, most farmers take risks that might endanger their health.

Some general information is provided below to illustrate the seriousness of gas, dust, and chemical problems. More detailed information can be found in the resources listed at the end of this chapter.

DUST AND MOLDS

Farmer's lung (also known as extrinsic allergic alveolitis or hypersensitivity pneumonitis) is an incurable, allergic

Dust and mold spores are encountered in many agricultural activities and are often associated with respiratory illnesses.

lung disease caused by the inhalation of spores found in moldy crops, such as hay, straw, corn, silage, grain, and tobacco; or in bird-breeding and mushroom-growing operations. A milder response from exposure to relatively low levels of dust is marked by coughing, shortness of breath, sweating, sore throat, headache, and nausea. An acute attack triggered by heavy exposure to dust starts with fever, muscular aches, and a general unwell feeling or malaise. These symptoms are accompanied by tightness in the chest, a dry cough, and shortness of breath. A chronic response develops after persistent acute attacks and recurring milder responses. It is marked by increasing shortness of breath, occasional fever, loss of weight, and general lack of energy. The victim suffers permanent lung damage, and in the worst cases death may occur.

Q fever is an infectious disease that spreads from animals to humans. Farm animals such as cattle, sheep, and goats can carry the Q fever microbe in tissues involved in birth—uterus, placenta, and birth fluids. Infected animals also release the microbe in milk and manure. People acquire the infection by inhaling infectious aerosols and contaminated dusts generated by animals or animal products. This microbe can survive for months and even years in dust or soil. Common symptoms resemble a serious case of the flu with high fever, chills, and sweating. In some cases, people develop liver and heart disease.

Organic toxic dust syndrome (also known as pulmonary mycotoxicosis, silo unloader's syndrome [SUS], or atypical farmer's lung) is a flulike illness due to the inhalation of grain dust with symptoms including fever, chest tightness, coughing, and muscle aches. The inhalation of the grain dust may occur in an agricultural setting or from covering a floor with straw.

GASES

A variety of potentially toxic gases are produced during many routine agricultural operations. When silage is put into a silo, plastic wrapper, or covered silage bunk, natural chemical changes give off gaseous byproducts, including nitrogen dioxide. A yellow, brown, reddish, or orange cloud or a bleachlike odor indicates a high concentration. When this gas comes in contact with moisture in your lungs, it can form nitric acid strong enough to severely burn the inside of the lungs and result in injury or death. Keep children away from silage storage areas. Anyone who goes near a silo must learn about the dangers of silo gas and be trained in the proper safety procedures.

CASE STUDY: DANCING UP A FEVER

An outbreak of organic dust toxic syndrome occurred at a college fraternity within 1.3 to 13 hours of attendance at a party where there was a dense airborne dust from straw that had been laid on the floor. The duration of illnesses ranged from 4.5 hours to 7 days. Symptoms were muscle aches, coughing, and a low-grade fever. Of the 67 fraternity members who attended the party and answered a questionnaire, 55 became ill (attack rate of 82 percent). The risk of illness was higher for those who spent more time at the party. Features of this outbreak were characteristic of organic dust toxic syndrome, an acute respiratory illness caused by inhalation of molds growing on hay, silage, or other agricultural products.
—from the Journal of the American Medical Association, W. T. Brinton et. al., Vol. 258, no. 9, September 4, 1987.

As manure rots, several hazardous gases are naturally formed. In the manure piles typical on small farms, these are not likely to be a problem because they are usually not confined, but in a liquid manure holding system the gases are trapped in small bubbles and are released when the manure is agitated or pumped. If larger manure handling is part of your farming operation, consult local detailed resources on safety precautions and ways to help minimize gas formation.

General information is provided below to illustrate the seriousness of the gas problem. Plenty of detailed information can be found in the resources listed at the end of this chapter. If you see any warning signs that display symbols of the gases listed below, keep clear and warn others.

Hydrogen sulfide (H_2S) is a clear, colorless gas and is by far the deadliest of manure gases. It is heavier than air and tends to pool in low areas. H_2S is known for its rotten-egg odor, but never rely on the odor or lack of it for a warning because high concentrations of this gas temporarily paralyze the nerves in your nose so you are unable to smell it. Breathing even small amounts of H_2S can result in nausea, coughing, headache, dizziness, and eye irritation. High concentrations of H_2S cause immediate stoppage of breathing through paralysis of the diaphragm.

Carbon dioxide (CO_2) is a gas produced by manure bacteria as a product of metabolism. In manure pits, CO_2 produced by the bacteria in the manure may displace the oxygen and cause difficulty breathing.

Ammonia (NH_3) is a colorless, lighter-than-air gas. It is easy to detect because of its sharp odor. High amounts (greater than 150 parts per million) of NH_3 can cause harsh coughing and severe irritation of the throat, eyes, and lungs. If the amounts are high enough, suffocation may result.

Methane (CH_4), also known as swamp gas, is a colorless, odorless, lighter-than-air gas that is extremely flammable and easily explodes. A spark or lighted match dropped into a manure tank or pit can have deadly results. CH_4 can also displace oxygen and cause suffocation.

Carbon monoxide (CO) is a clear, odorless, and very deadly gas formed during combustion in equipment, such as internal-combustion engines and space heaters. If you are operating an older, less well-maintained farm, watch out for areas where CO may collect, such as in shops with old heaters, generator shacks, or poorly ventilated grain/feed bins near augers powered by gasoline engines.

Paints and solvents can also give off various hazardous and/or flammable gases as they dry, so be sure to work with proper protective equipment and plenty of ventilation when using these materials.

Motor fuels and lubricants present another potential soil and water contamination hazard if they aren't stored and transferred with the proper procedures.

Most rural areas now have approved hazardous material disposal sites. To locate the one in your area, ask your local fire department or chemical dealers.

CHEMICALS

Many different chemicals may be encountered on a farm, even if no pesticides are purchased or used. Proper use, storage, and disposal of chemicals protect you and others from accidental exposure. Proper storage and disposal are particularly important for protecting inquisitive children exploring the farm. Proper handling is also vital to preventing injury to farm animals, water, and the environment.

Whether you're using farm chemicals or moving them to a proper disposal site, it's important to follow the manufacturer's recommended handling and application procedures and wear correct personal protective gear, such as appropriate respirators, eye protection, impermeable aprons, and gloves.

If the chemical is purchased new, it generally comes with a copy of the Material Safety Data Sheet (MSDS). The MSDS outlines any health hazards and appropriate precautions to take, along with other important information like handling and disposal procedures. For materials without an MSDS, visit the manufacturer's website or visit these Internet sites:

Agricultural chemicals:
Crop Data Management System Inc. at www.cdms.net
Household products:
National Library of Health product database at
www.ehso.com/msds.php

FARM DUST, MOLD, GAS, AND CHEMICAL INFORMATION

Farm and rural area safety information is a key mission of the continental network of agricultural extension services at land grant universities and government research stations. Your local branch can be found in the telephone directory or at these sites:

State Departments of Agriculture:
www.state.sd.us/doa/department
Canadian Provincial & Territorial
Departments of Agriculture:
www.agr.gc.ca/province_e.phtml

SOME OF THE MANY SAFETY ADVISORIES AVAILABLE ARE:

Dust and mold
www.cdc/gpv/nasd/docs/d000701-d000800/d000738/d000738.html
Farmer's lung
www.cdc.gov/nasc/docs/d001601-d001700/d001609/d001609.html
Organic toxic dust
www.gov.on.ca/OMAFRA/english/livestocks/swine/facts/93-003.htm
Q Fever
www.uaex.edu/biosecurity/cross_referenced/QFever_questions_answers.asp
Gases
Silo gas dangers:
www.cdc.gov/nasd/docs/d001601-d001700/d001621/d001621.html
Manure gas dangers:
www.cdc.gov/nasd/docs/d001601-d001700/d001616/d001616.html
Chemicals: *Chemical/pesticide topics:*
www.cdc.gov/nasd/menu/topic/chem_general.html

CHAPTER 7
SAFETY AROUND LIVESTOCK

▶ *Purdue University Extension statistics indicate that one in six farm injuries involves animals, while the National Agricultural Safety Database statistics show the rate to be as high as one in three.*

▶ *Women and girls have a horse-related injury rate more than twice the rate of men, who in turn suffer a majority of the cattle-related injuries.*

▶ *Over 40 percent of the on-farm animal-related injuries occur among children under 10 years old.*

One of the attractions of farm living is the opportunity to raise some cows and calves, a few sheep, horses, or other animals. Over time, these animals often seem like a pet or part of the family, especially to children and visitors. But all livestock should be considered unpredictable. They have important instinctive behavior patterns such as rejoining the herd, maintaining herd hierarchy, maternal protection of newborns, male attack of intruders, and reaction to sudden movement and noises beyond the range of human hearing. You may recall cowboy movies where a subtle noise stampedes the herd, and that is not much of an exaggeration. On a smaller scale, moving or flapping objects can also disrupt handling as you try to move a few head of cattle from one pasture to another. A cloth or coat swinging in the wind or turning fan blades can cause animals to balk or run wildly. Movement at the end of a chute can cause them to refuse to be herded.

Most farm animals are herd animals, and separation from the herd causes a lot of anxiety. The result can be panicky or aggressive behavior toward herders.

Consider the weight of the animal and what that could mean if all that mass collides with you. A 1,200-pound horse or 1,500-pound bull pressing you against a fence is bad enough, but if that animal is also moving at the time, that's lots of kinetic energy that gets absorbed by fragile human bone and tissue. Even a 200-pound pig, moving at top speed and accelerating amazingly fast, can knock you off your feet when you're trying to catch one for veterinary treatment.

Keys to keeping you and your family safe around livestock include having knowledge about instinctive reactions of various animals and how to use that knowledge to advantage when herding, feeding, treating, or handling livestock. You also need to know how to keep livestock areas free of hazards, such as improperly designed chutes and lanes, lack of human escape routes, and sharp protrusions.

Danger signs in livestock:
• Raised or flattened ears.
• Raised or rapidly lashing tail.
• Raised hair on the back.
• Bared teeth.
• Rolling eyes.
• Making unusual or distressed cries.
• Climbing onto the backs of other animals to escape.
• Snorting, tossing the head, and pawing the ground.
• Stiff-legged gait or posture.
• Any history of previous aggression.

Be especially cautious around animals such as this blind cow, who is normally a very well-tempered animal. But like all blind or deaf animals, she can suddenly swing around to investigate disturbances with other remaining senses such as smell. If you are standing too close, you could easily be bruised, knocked down, gored, or trampled.

CASE STUDY: CRUSHING BY ANIMALS

1) A 48-year-old dairy farmer was helping load cattle onto a trailer when he was fatally crushed between the end of a gate and a steel fence. The farmer and two other workers were attempting to load cows onto a trailer using a chute created by fencing and portable gates. One of the cows turned and rushed back through the gate into the barn area. The workers were able to turn the cow so that it was once again going toward the trailer and had passed through a makeshift gate. Once again the cow turned and tried to push through the gate while the victim was standing by the side wall at the open end of the gate. The victim was crushed by the gate and his heart was punctured by a metal protrusion on the end of the gate.
—from New York State FACE Report #03NY040

2) Two elderly farmers were found dead in the pen area of a young Black Angus bull on the farm where they had lived and worked for the last 27 years. The men had multiple rib fractures, apparently from getting butted up against the side of the nearby barn. The two victims were found by a third brother, who had just returned from the hospital, where he had been recuperating from injuries suffered when attacked by the same bull five days earlier.
—from Iowa FACE Report #00IA055

Male animals, such as bulls, are naturally aggressive, and their extra bulk and fighting tools (horns, spurs, etc.) makes even a small collision dangerous to handlers. Bulls, rams, stallions, roosters, and other male animals may become especially excited and aggressive when they detect breeding females nearby.

When an animal being driven down an alley suddenly turns to look at the herder may think of making a dash backward. Be prepared to get out of the way fast.

LIVESTOCK WITH YOUNG

Mothers of newborn animals can exhibit a very strong maternal instinct that includes being very defensive of their young, and therefore can be difficult to handle if they must be moved from place to place. When possible, let the young stay close to the adult during handling.

VETERINARY PROCEDURES

Taking proper care of animals' health may involve procedures such as cleaning, administering medicines, treating wounds, or trimming overgrown hooves. It's no surprise that most animals take exception to these unfamiliar procedures, even though you mean well. When a large animal expresses its displeasure by thrashing wildly about, even if they are confined in a squeeze, there's a very serious risk of injury from crushing, kicking, goring, or other causes.

HEALTH AND HYGIENE

Good hygiene is vital to human safety in livestock management, especially in confined areas. Good ventilation and sanitation minimize problems from dust and molds that can cause respiratory and digestive problems, such as farmer's lung or organic toxic dust syndrome. For more details on these problems, see Chapter 5.

The maternal instinct for a mother defending her young can also create an extreme safety hazard as she attempts to butt, gore, kick, bite, or trample anyone coming near her young.

Good hygiene helps limit the livestock-to-human transmission of diseases and organisms such as fleas, Q fever, leptospirosis, brucellosis, tuberculosis, salmonellosis, ringworm, or anthrax. Wear rubber gloves, as well as other protective clothing, when working with sick and injured animals. Practice personal hygiene by washing your hands and face after handling animals.

DEALING WITH ANIMAL DANGER SITUATIONS

These examples are not a complete list and are provided only to show the kind of instinctive behaviors related to safety. Consult other owners and your local agricultural extension services for information on handling typical farm animals such as horses, cattle, pigs, sheep, and goats. Most animals have a strong territorial instinct and develop a very distinctive, comfortable attachment to areas, such as pastures and buildings, water troughs, worn paths, and feed bunks. Forcible removal from these areas may cause animals to react unexpectedly.

Large animals can generally see at wide angles around them, but there is a blind spot directly behind their hindquarters. Any movement in this blind spot will make the animal uneasy and nervous. The sense of smell is extremely important to animals, especially between females and newborns. Often animals react to odors we do not detect, such as nearby water or predators. Handling facilities should be only one color since all species of livestock are likely to balk at a sudden change in color or texture. Baby male animals

Preparing to administer medication to an animal can be a dangerous time because the animal is generally quite anxious about being separated and handled. It may decide to bolt while your attention is focused on handling the medicine or administering the dose.

For exotic animals such as llamas, emus, ostrich, bison, and others, consult with experienced owners on safe handling and how to deal with unique instinctive behaviors.

that are bottle-raised will, on reaching sexual maturity, be aggressive toward humans. Be wary around these animals.

CATTLE, PIGS, AND SHEEP

- These herd animals are calmed by being in groups. Separation from the group can lead to a panicky flight behavior. Conversely, the herd can be moved by concentrating on the dominant animal so the others will follow.
- Their instinctive tendency is to move from a dimly lit area to a more brightly lit area, provided the light does not hit them directly in the eyes. A spotlight directed on the ramp will often help keep the animals moving.
- Always announce your presence when approaching. For cows, lightly touch the animal rather than making the first contact with a bump, shove, or a poke with a stick.

- Those dairy cows may look content in the pasture, but they are generally more nervous than other animals. They are easily startled, especially by strange noises and people.
- Herd pigs with a lightweight panel to prevent them from trying to make a sudden dash past you. Quietly and gently make yourself known to avoid startling the pigs. A knock on the door or rattling the door handle will usually suffice.
- Sheep have difficulty picking out small details, such as the open space created by a partially opened gate.
- Always keep a close eye on rams (male sheep) that often butt if your back is turned.
- A sheep can be immobilized for safe handling by sitting it upright on its rump.
- Small pigs can be handled safely by grabbing the back legs and raising the animal off the ground.
- Cows and pigs have poor depth perception and difficulty in judging distances. For this reason cows are

All livestock tend to refuse to walk over any change in flooring texture or surface such as a drain, grate, hose, puddle, or shadow. For this reason, be prepared for the animals to be hard to move from a loading chute into the truck.

apparently unable to distinguish the difference between a real cattle guard and parallel stripes painted on a road.

HORSES

- Ears point toward where attention is focused. Ears that are flattened backward warn you that the horse is getting ready to kick or bite.
- Always work with calm and deliberate movements around horses. Nervous handlers can make horses nervous, which creates unsafe situations.
- Be careful when approaching a horse that is preoccupied, such as when its head is in a hay manger. Speak to the horse to get its attention and wait until it turns and faces you before entering the stall.

- Always use a lead line to prevent getting a hand caught in the halter if the horse unexpectedly moves.
- To lead a horse through a doorway, step through first, then quickly step to the side and out of the horse's way.
- Never wrap any piece of tack around your hand because it could tighten and injure the hand if there is sudden movement.
- After you remove the halter, make the horse stand quietly for several seconds before completely letting it go. This will help prevent the horse from developing a habit of bolting away and kicking at you.
- Do not climb over the lead line of a tied horse. The horse may pull back and jerk the line tight, which is especially bad for men!

CHAPTER 8
FARM FIRE SAFETY

▶ *In the United States, losses from fire accounted for 37 percent of farm equipment losses and 46.5 percent of reported livestock loss, according to a five-year study by the American Association of Insurance Services.*

▶ *Every year, fire destroys 1,700 farm buildings and 66,000 areas of grassland on the (approximately) 300,000 large and small farms in the United Kingdom. Fifty percent of these fires are started deliberately, either as an act of vandalism or a fraudulent insurance claim.*

Getting away from it all by moving to a small farm can also mean getting farther away from emergency services, such as firefighters. Even where there is rural volunteer or professional firefighting services, factors such as longer travel distances may increase response time. If your access road is not clearly marked or hard to use for large vehicles, this may add to the difficulty of firefighters finding your farm.

In the country, firefighters cannot hook up to a fire hydrant, nor is it always practical to carry the large volume of water necessary for combating a large fire, such as in a barn or shop.

FIRE READINESS

• Have simple discussions so everyone knows what to do or where to go in case of a fire. It is especially important to protect children, so you and firefighters don't spend critical moments searching for them.

• Human safety, including your own life, must be your first priority. Make sure you, your family members, and employees are safe.

• Call the fire department immediately and let the experts take control. People have been seriously injured or killed when trying to save animals, grain, or equipment, forgetting that smoke and toxic fumes can kill them in seconds.

• Have a list of what flammable items are stored where so firefighters know what to be prepared for and what to protect themselves against. Location of fuels, lubricants, and pesticides is especially important.

• Firefighters may ask you for priorities on which building to save. Have a plan on whether it's livestock, machinery, or feed. If a livestock building is on fire, animals may already have been exposed to deadly heat, smoke, and gases. It may be safer and more realistic to save an adjacent building or vehicles stored in it.

FARM ACCESS

• Be ready to give the fire dispatcher clear, concise directions to your farm—no "turn left where the Smiths' barn used to be" or similar confusing terms. Add GPS coordinates if you can provide them.

- Make sure that the access roads or trails to farm areas are smooth and can support large vehicles. Many farms have only a rutted, bumpy trail to the yard, which makes it difficult to haul firefighting equipment in and out. If a truck bringing gravel to your yard has a hard time getting in and turning around, firefighting equipment can have as much or more trouble.
- There should be a reliable, accessible water source, such as a pond, no farther than 100 feet from major buildings. Fires can occur in winter, so don't count on seasonal water flows or those that ice up heavily.

LIVESTOCK BEHAVIOR DURING FIRES

- Animals may panic and refuse to leave a burning building, or in some cases they may run back in after being led out. Don't become trapped by repeated attempts to evacuate them. Cows and horses tend to panic if they are frightened or forced to use an unfamiliar exit.
- If you are able to evacuate animals, be sure you are not leading them toward a dead end, such as a gate that won't open outward. What may ensue is panicky flight of the animals, with the possibility of you or others being trampled or crushed.
- If animals have suffered from heat, smoke inhalation, or burns, immediately call a veterinarian to examine and treat them. If possible, lightly spray water on animals to cool them and reduce shock.

Hay (loose or baled) that is stored at too high a level of moisture can start to heat up and eventually burn. Make certain that all hay is properly dried before you put it in a building or outdoor stack. Cover outdoor stacks to prevent infiltration of rain that can lead to mold and subsequent heating.

HAY AND GRAIN STORAGE

- If hay is slowly smoldering in an upper level of a barn, call the fire department and begin evacuation if possible. Do not try to throw smoldering hay out a window or door because exposure to oxygen will make the fire flare up.
- If you see or smell smoke coming from the hay, place boards or plywood on the hay before walking on top of it to probe for hot spots. The boards or plywood will spread your weight over a larger area to prevent falls into burned-out cavities below the surface. A rope as a lifeline is also recommended.
- Bin dryers should be equipped with adequate controls that will automatically shut off heaters, blowers, or dampers when temperatures get too high.
- Dust resulting from grain movement poses a risk of explosion and fire. Ensure adequate ventilation and avoid sparks during grain handling.

CONSTRUCTION PLANNING

- When constructing any major farm building, the site plan should allow for adequate spacing between buildings to prevent the spread of a possible fire.
- Locate new buildings at least 40 feet away from above-ground fuel storage tanks to minimize the potential for spread of fire. Consider a longer distance if the buildings are in line with prevailing winds.
- When constructing a new building, include fire-prevention features such as fire doors, a firewall between hay/bedding storage and stabling or work areas, flame-retardant or fire-resistant materials, fire-retardant latex paint, smoke detectors, fire alarms, and automatic heat-sensitive sprinkler systems.

If you're working on an old farm, the wiring is likely to be substandard by today's standards. All wiring and fixtures should be inspected for exposed wires, broken insulators, improper grounding, or improper installation, and brought up to current codes.

Check around the yard so you know the location of electrical shutoffs that can cut off power to buildings or areas in case of a fire.

- Install battery-powered emergency lighting to permit evacuation of people and animals in case of power failure, and create a water source on or near the premises.

ELECTRICAL SAFETY
- Fixtures for fluorescent lights should have dust- and moisture-resistant covers. Incandescent bulbs should have globes sealed against moisture and dust, plus a metal cage to prevent accidental breakage. Older barns with lightbulbs hanging from a wire are a fire waiting to happen.
- Make sure the wiring and fuses/breakers can meet your current needs without overloading the electrical system. Overloads cause excess heating, which can lead to fire in the building.
- Use waterproof wiring and receptacles, enclosed electric motors, and similar equipment in any buildings that are cleaned with water.
- Locate the main electrical panel in the driest, most dust-free area. The panel housing should be corrosion resistant and weatherproofed even if it is installed inside a building.
- Wires (even temporary ones such as extension cords) should be run through conduit pipe to keep the wires safe from breakage and away from the chewing of horses, livestock, and rodents.

- Electric motors in livestock facilities should have moisture/dust-proof on/off switches. Motors should not be within 18 inches of any combustible material such as hay or bedding.
- Electric fence units can be potential fire hazards, especially if the units are of the continuous-current type, rather than the solid state type with intermittent current. Fence charging units should be 10 feet from buildings and enclosed in a weatherproof housing.
- All major farm buildings should be equipped with lightning rods or other suitable lightning grounding provisions.

KNOW YOUR FIRE EXTINGUISHERS
- Keep a fire extinguisher in every building where it's practical so that if a fire develops you don't spend crucial time running a distance to find an extinguisher.
- Make sure the fire extinguishers you buy match the type of fire that can be expected in your situation. If you use the wrong unit on a fast-moving fire, you could cause the fire to spread even faster. Read the label on the fire extinguisher and get advice from where it was purchased.
- Study how to use the extinguisher so you're ready before a fire ever starts.
- Make sure all extinguishers are serviced at intervals that don't exceed one year.
- Know your limits and always think "safety first." Fire extinguishers cannot do the job of a local fire department. When a fire burns for more than a couple of minutes, the heat starts to build up and intensify. Once this happens, you are past a point of using an extinguisher. Get out of the building and let firefighters handle it safely.

FARM FUEL AND LUBRICANTS
- Be sure all containers for flammable and combustible liquids are clearly and correctly marked so that they do not end up near sources of heat or sparks.
- Never store fuel in breakable food or drink containers.
- Watch out for and immediately take care of leaks or deterioration in fuel storage and delivery equipment.
- Before cutting, welding, or soldering a fuel tank, completely remove any vapor or liquid. Explosion and fire can result from using a torch on a tank believed to be empty.
- Do not keep gasoline inside the home or transport it in the trunks of automobiles or RVs. If gasoline must be transported, carry a small amount in a labeled

safety can on the floor of the back seat. Roll down the windows so moving air can sweep away vapors.

- Diesel, motor oil, and grease require more heat than gasoline to ignite, but once ignited they will burn long and hot. Keep them away from heat and sparks.

PORTABLE HEATERS AND HEAT LAMPS

- Portable heaters should not be used in the barn area. If heating units are used in tack rooms, they should not be left unattended because they may not have safety devices that prevent overheating.
- Make sure heaters have a shutoff device that activates if the unit is knocked over. Place the unit where livestock cannot knock it over.
- Heat tapes and water tank heaters must have a thermostat and be adequately protected so that horses, livestock, cats, dogs, or rodents cannot chew them. Chewing can result in electrocution, shocks, and/or fire.
- Heat tapes should be protected with a fire-retardant insulation material.

GENERAL FIRE SAFETY

- Smoking should never be permitted in any building where combustible materials are located or stored.
- Exit doors should be clearly marked.
- Aisles should be raked or swept clean of hay and bedding. Vacuum cobwebs and dust regularly. Wipe dust/dirt off light fixtures, outlet covers, switches, and panel boxes.
- Keep weeds, dry grass, and trash away from the outside of the barn.
- Be very careful about burning trash, especially in windy or dry conditions. For burning large amounts of trash or crop materials, obtain a fire permit and notify the local fire department to prevent false alarms.

Many fires are accidentally started when the wind stirs up a burning barrel. Keep a screen atop the barrel when it is left burning unattended.

Middle, right: *When welding or grinding metal in the workshop, make sure that any sparks and hot metal will not fly toward combustible or explosive materials. Keep a supply of suitable heat-quenching liquid nearby, as well as a fire extinguisher.*

Right: *Driving or parking in dry grass and stubble can start a fire from the heat of the vehicle exhaust. This is especially a hazard on vehicles equipped with catalytic converters, which operate at very high temperatures. Make sure the vehicle is equipped with a working spark-arresting muffler and avoid driving or parking in areas where combustible materials can touch the bottom of the vehicle.*

CHAPTER 9
CHILDREN'S SAFETY

▶ *The Centers for Disease Control and the National Institute for Occupational Safety and Health (CDC/NIOSH) reported in July 2001 that drowning was the second-leading cause of death on farms (27 percent), with children under five years of age accounting for 32 percent of the deaths.*

▶ *Drowning is the leading cause of farm deaths in 17 states. Children under the age of four are at highest risk for drowning in farm ponds.*

A longstanding attraction of rural living is the perceived chance for your children, grandchildren, or their friends to roam and play freely in the country and encounter nature firsthand. However, there are key safety aspects to consider in order to prevent them from being injured or killed by drowning, machinery injuries, animal encounters, falls, or other hazards.

While you can never "childproof" a farm and can't be hovering over them every second like an anxious hen with chicks, attention to some of the prime accident areas can help eliminate a lot of needless risk. In many cases, the changes you make are also beneficial, even if children aren't around. For example, fencing a farm pond keeps children safer and reduces the potential for escaped livestock to pollute your water source with urine, feces, and drowned carcasses.

Key areas and precautions are listed below. Children are avid explorers, which often means they are discovering new ways to get hurt, but there are general prime injury areas.

WATER BODIES

- Supervise children at all times around water, just as you would at the pool or beach.
- Wherever practical, fence in all ponds and lagoons with self-closing, self-latching gates as part of the fence. Post the fenced area with "no trespassing" signs and explain to children what the signs mean.
- Where swimming or wading is allowed (or perhaps not allowed, but suspected to occur!), make rescue tools available, such as a rope secured at one end to a sturdy post and a floating weight at the throwing end, a flotation ring, a long pole, and/or a loud whistle to summon help. Explain and demonstrate the use of the safety tools to children.

To avoid putting any actual children at risk in images that illustrate particular risks, Rascal the Safety Monkey has been used instead.

That is not a good place to ride, Rascal! Children have fallen off tractor fenders and been run over by the implement or wagon towed behind the tractor.

Canals, ponds, streams, lagoons, and other bodies of water are endlessly fascinating for children, for such activities as fishing; watching bugs, frogs, and minnows; skating; or throwing rocks in the water. But steep banks, strong currents, or deep mud in farm water bodies can turn play areas into death traps.

- If a dock or fishing platform is located at a pond, install a ladder that extends into the water. Children may not be strong enough to climb out of the water and onto a slippery dock, especially if they are scared or in shock from cold water.
- If you live in an area where water bodies freeze over, explain to children, at their level of understanding, basic ice hazards and ice rescue. Your local school, library, Red Cross unit, or Scout/Guides chapter can provide this information.
- Irrigation canals should be considered off-limits to both children and adults, especially in main canals. The current is powerful and steep banks make it extremely difficult to climb out of the water.
- Warn children to stay far away from irrigation wells, spillways, gates, intakes, motors, and other system parts. Explain to children about the strong currents and other hazards that make it imperative to keep away. Install proper safety shielding on all irrigation mechanisms.

That's not a good place to ride, either, Rascal! Children have fallen off the hood and been run over by the tractor's big back wheels.

MACHINERY

Farm machines, big or small, can be fascinating to children, but small children should not be around machinery without close supervision and secure seating. Fenders or laps are not good enough. Youths learning to operate machinery also need a secure trainee platform and careful training, even in activities you think might be taken for granted, such as operation of tractors at road speeds.

119

Stay away from there, Rascal! The driver can't see you, and you could be squished by the wheels as the driver backs up.

ATVs are very useful around the farm. Slow operating speeds around the farm make it seem okay to ride, but don't do it, Rascal.

Here's why children shouldn't ride on ATVs. Falling off is bad enough, but getting run over is also a possibility.

Rascal likes playing with his truck and pretending he's a grownup truck driver, but he really should find a better place to play. This also shows why it's an excellent idea to have backup alarms on all farm vehicles.

Never allow children to ride in the bucket of a tractor's front-end loader. A sudden jolt can bounce them out of the bucket and put them at extreme risk of being run over. Even if they don't fall out, the shock of landing back in the hard steel bucket can cause tissue or bone injury.

PRACTICAL SAFETY PRECAUTIONS AROUND FARM MACHINERY

Accident records make it all too clear that it's really easy for children and youths to be thrown from, entangled in, crushed by, or mangled by tractors, implements, threshers, lawn mowers, trucks, and all-terrain vehicles (ATVs). That's why a few simple, practical precautions should be part of your day when children are around.

- Get a definite, visual confirmation of where children are before moving a tractor, implement, or other piece of equipment. Unless you can actually see that they are in a safe position, small children may be out of sight behind wheels, tracks, or other machine parts.
- Never allow riders on a tractor, implement, harvester, ATV, or other piece of equipment unless a secure operator's seat or platform is provided. A child that falls off due to a sudden jolt or turn is at extreme risk of being run over or entangled. Due to the size of the machine, the operator may not even see the accident in time to prevent it.
- Children riding on fenders or laps may interfere with the operator's access to machine controls and increase the risk of an accident.
- Never allow riders in the back of a pickup truck without seats and grab rails. A sudden stop, turn, or jolt may smash the rider forward into the rear window or throw them out of the truck.
- When dismounting from a tractor to adjust or repair an implement, make sure all riders dismount and remain a safe distance from the machine. Unexpected accidental engagement of the implement drive by a rider may cause injury or death to the operator.
- Take the time to continually explain what you are doing so children do not have to satisfy their curiosity about the job by getting close to hazardous machinery parts.

Rascal is exploring the machinery workshop, as all children like to do. Take time to explain the dangers and which tools are off-limits.

Left: Rascal is lighting straws to see how long they can burn before he pinches them out. It's a practice that seems to fascinate every generation of farm kids, so the full extent of the risks should be explained before they cause a large fire.

FALLS
Farm fall prevention for children

- When not using a ladder, lay it down on the ground or secure it on a wall storage rack. Besides being a safety benefit, this prevents ladder damage from getting knocked down by wind or animals.

- On machines such as harvesters with elevated access platforms, keep platforms clear of mud, tools, twine, or other obstructions that could lead to slips and falls.

- Keep the area around haystacks clear of protruding stakes or other objects that could impale a falling person.

Above: *Many serious falls occur because children are fascinated by climbing up onto high places such as sheds, haystacks, machinery, and fuel tanks.*

Upper right: *All working wells and cisterns should be securely covered with a strong lid too heavy for a child to lift. Fill in any abandoned wells, cisterns, and other excavations.*

Middle right: *Keep the lower ends of permanent access ladders to high areas, such as silo tops, above reach for children.*

Right: *Rascal is playing around a grain bin where the ladder extends down near the ground.*

Advise children to feed animals from outside the pen whenever possible. In a simple rush to get to the feed, the animals may knock children down.

Rascal decides to check out what's on top of the ladder, not realizing it's a really long way to the bottom if he slips.

FARM ANIMALS

A look through any selection of children's books shows most children are fascinated by animals, and they may assume farm livestock are pets to play with. Baby animals are even more irresistible. However, all farm animals should be assumed to be unpredictable and must be treated with respect and caution. The large mass of a farm animal, such as a cow or horse, means accidental or aggressive contact with a small child will result in extensive injury. Most children do not have sufficient strength or speed to quickly scramble out of the animal's pen in a dangerous situation.

Children love animals, but they often do not have the experience to recognize the subtle signs an animal will exhibit just before defending itself, its territory, or its young.

PRECAUTIONS FOR CHILDREN NEAR FARM ANIMALS

- Move slowly and avoid sudden movements, screaming, or running.
- Always approach animals from the front so they can see what you are doing.
- Wherever possible, approach in the company of someone already familiar to the animal. Seeing someone they recognize reduces stress in the animal.
- A mother animal is usually very protective of her young and may vigorously defend them against anyone she sees coming close.
- Avoid contact with bulls, boars, rams, and other male animals. They tend to be naturally aggressive.

WHERE TO GET MORE INFORMATION

Your local agricultural extension office can provide a wealth of practical information on keeping kids safe on farms. An Internet search using a term such as "farm safety children" will also provide a great deal of useful advice, such as:

National Ag Safety Database: Child Safety:
www.cdc.gov/nasd/menu/topic/child_safety.html

Farm Safety for Children: What Parents and Grandparents Should Know
ohioline.osu.edu/aex-fact/0991.html

Keeping Farm Children Safe:
Risks and precautions at Various Developmental Stages
www.extension.umn.edu/distribution/youthdevelopment/DA6188.html

North American Guidelines for Children's Agricultural Tasks
www.nagcat.org/nagcat/pages/default.aspx

CHAPTER 10
LIFE AFTER
A FARM INJURY

Despite one's best efforts at farm safety, accidents can still happen and leave a surviving victim temporarily or permanently disabled. In that case, it's not necessarily the end of life on the land. Many farmers and their families find successful ways to keep living and working on their farm. However, it is a formidable challenge for both the injured farmer and spouse or other caregiver. Barriers faced by farmers and ranchers with disabilities can include:

- Lack of information about effective home and worksite modifications.
- Lack of money to pay for necessary modifications.
- Economic difficulties from lack of income and loss of insurance.
- Long distances from needed services.
- Scarcity of professionals trained in helping people accommodate their disabilities in an agricultural occupation.
- Attitudes among others about the ability of a farmer with disabilities to continue in this high-risk, physically demanding occupation.

BOB GUEST

Bob Guest was four years old when he became entangled in a PTO and lost his right arm above the shoulder. Bob raises cattle and horses, and as a member of Farmers with Disabilities, he supports other farmers who have suffered disabling injuries. He also visits schools and talks to children about farm safety

"I was allowed to put the pillow on my dad's tractor seat. It was an old tractor with a steel seat. I would take it off at night and bring it in to stay dry and I'd put it back in the morning. Dad was welding when I saw the pillow fall off. I went to put it back on, and I was just the very height of the PTO, which was turning. The pillow twisted up in the power takeoff and it cut off my arm."

In addition to those barriers related to returning to farming despite disabilities, many farmers and ranchers are also at risk of acquiring secondary injuries or secondary conditions. The Rural Research and Training Center at the University of Montana states that the average person with a disability reports 14 secondary conditions. These include injuries to shoulders, wrists, backs, and other body parts.

While the challenge cannot be overstated, meeting tough challenges is part of farming. To help work through the ongoing challenges, there are now organizations and resources for farmers with disabilities who make the choice to keep living and working on their farms.

EARLY ORGANIZATIONS IN THE UNITED STATES

One of the first organizations in the United States to help disabled farmers was the Vermont Rural and Farm Family Vocational Rehabilitation Program that was established in 1966. This program was a partnership between the University of Vermont Cooperative Extension Service and the state office of Vocational Rehabilitation.

In 1984, Purdue University established a facility to respond to farmers and ranchers throughout the country who wanted information and technical assistance on equipment modifications, resources, and ideas related to farming or ranching with a physical disability. A grant from the National Institute on Disability and Rehabilitation Research was key to creating this center. Breaking New Ground has since distributed resources and publications worldwide.

Also in 1984, the University of North Dakota and Memorial Hospital in Grand Forks began a partnership to provide specialized rehabilitation for farm-accident injuries. Sponsorship of two national conferences on rural rehabilitation technology brought national attention to this much-needed program.

In 1986, The Easter Seal Society of Iowa (now called Easter Seals Iowa) established the Farm Family Rehabilitation Management (FaRM) Program. This program provides onsite rehabilitation and assistive technology services to Iowa farm families affected by disabilities. In 1989, the FaRM program received the national Program Innovation Award from the National Easter Seal Society (now called Easter Seals) and a commitment from the national organization to help other Easter Seals affiliates throughout the country to develop similar programs.

AMERICAN NATIONAL PROGRAM ESTABLISHED

Based on the success of the Vermont, Indiana, North Dakota, and Iowa projects, Easter Seals took the idea of establishing a national program to Congress. Senator Tom Harkin from Iowa, Senator Patrick Leahy from Vermont, and the late Senator Quentin Burdick from North Dakota sponsored federal legislation to establish what is now called AgrAbility.

As part of the 1990 Farm Bill, the AgrAbility Project was authorized to provide education and assistance to agricultural workers with all types of disabilities and their families through unique partnerships between the USDA Cooperative Extension System and private nonprofit disability service organizations. AgrAbility Projects in 24 states have now assisted more than 10,000 farmers affected by physical, sensory, cognitive, or emotional disabilities. AgrAbility program services include onsite technical assistance on worksite and home modifications to accommodate disability, education to prevent further injury and disability, training for extension educators and rural professionals to upgrade their skills in assisting farmers with disabilities and development, and coordination of peer support networks. The national project is now a partnership between the University of Wisconsin-Cooperative Extension Biological System Engineering Department and the Easter Seals national headquarters, and offers the following:

- Professional training and technical resources on agricultural worksite assessments.
- Identification of solutions to overcome barriers.
- Strategies and technologies for farming with specific limitations.
- Prevention of secondary injuries or illnesses.
- Alternative agriculture ventures.
- Funding ideas for needed technologies.
- Method, materials, and resources for providing effective services.

Currently, 18 states have AgrAbility programs. Each program is a partnership between the extension service at a land-grant university and a private, nonprofit, disability-related service provider. In addition to its own high-quality staff, each state project develops a network of skilled individuals and related organizations to which they can refer individuals or contact for assistance. These resources include fabricators, engineers, health and rehabilitation professionals, durable medical equipment dealers, agricultural equipment suppliers, educators, carpenters, and welders.

Many states also rely on the use of ingenuity networks comprised of volunteers with a variety of skills who are willing to assist individuals in obtaining or fabricating a needed solution. Agencies such as state vocational rehabilitation, state assistive technology projects, assistive technology exploration centers, and independent living centers frequently collaborate with state AgrAbility projects to provide quality services. Cooperation and collaboration is the key to success with all AgrAbility services.

CANADIAN PROGRAMS

The Farmers with Disabilities program began in 1985. Through the program, farmers across Saskatchewan formed a network they can rely on for advice and support. Injured farmers and farmers with disabling health conditions can talk with others who have experienced the same challenge. Farmers having trouble adapting their equipment for their disability can draw on the network for ideas.

Farmers with Disabilities publishes a newsletter called the *Handifarmer* that is full of news on upcoming conferences, ideas on adapting equipment, tips on safety and accident prevention, and stories of how other farmers are dealing with their disabilities. Recent issues of *Handifarmer* are available online. An extensive collection of resource material is available on the special tools, farming aids, and modifications or adaptations that have been made to farm equipment and yards by families who have already gone through the process. *The Changing Gear* machinery Modifications catalog is available online or on CD-ROM.

The Canadian Farmers with Disabilities Registry (CFWDR) was formed in 1997. It supports disabled people in pursuing farming as a viable and realistic occupational goal and lifestyle.

Farmers with Disabilities in Manitoba began in 1983. It struggled for many years, then was revived in 1996 when a farmer was in need of support and information about coping with a missing limb. The program now offers a variety of support and information services. Members visit accident victims in the hospital and offer support to both the victim and the family. Members also make public and private presentations on how to prevent farm accidents and deal with the aftermath of injuries.

FOR MORE INFORMATION

AgrAbility/Breaking New Ground
1146 ABE Building
West Lafayette, IN 47906
Phone: (765) 494-5088
Toll-Free: (800) 825-4264
E-mail: bng@ecn.purdue.edu
Internet: www.agrability.org

Farmers with Disabilities
310 Louise Avenue
Saskatoon, SK S7J 2C7
Phone: (306) 374-4448
Fax: (306) 373-2665
E-mail: farmerswithdisabilities@abilitiescouncil.sk.ca
Internet: www.abilitiescouncil.sk.ca/main/html/services/farmers

Canadian Farmers With Disabilities Registry (CFWDR)
R. R. 1
Aylesford, NS, B0P 1C0
Phone/Fax: (902) 847-9420
E-mail: info@fwdcanada.com
Internet: www.fwdcanada.com

Farmers with Disabilities in Manitoba
135 Elm Park Road
Elm Creek, MB R0G 0N0
Phone: (204) 436-3181 office hours,
(204) 436-2554 after hours
E-mail: info@fwdmanitoba.ca (please put FWDM2004 in "subject")
Internet: www.fwdmanitoba.com

INDEX

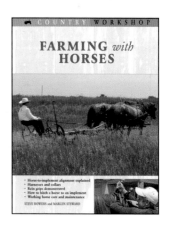